HOUNDS
WHO HEAL
It's a kind of Magic!

CHRIS KENT
Foreword by Dr Daniel Allen

Hubble & Hattie

The Hubble & Hattie imprint was launched in 2009 and is named in memory of two very special Westie sisters owned by Veloce's proprietors. Since the first book, many more have been added to the list, all with the same underlying objective: to be of real benefit to the species they cover, at the same time promoting compassion, understanding and respect between all animals (including human ones!) All Hubble & Hattie publications offer ethical, high quality content and presentation, plus great value for money.

More great books from Hubble & Hattie –

www.hubbleandhattie.com

First published October 2017 by Veloce Publishing Limited, Veloce House, Parkway Farm Business Park, Middle Farm Way, Pound Dorchester, Dorset, DT1 3AR, England. Fax 01305 250479/email info@hubbleandhattie.com/web www.hubbleandhattie ISBN: 978-1-845849-73-3 UPC: 6-36847-04973-7 ©Chris Kent & Veloce Publishing Ltd 2017. All rights reserved. With exception of quoting brief passages for the purpose of review, no part of this publication may be recorded, reproduced or transm by any means, including photocopying, without the written permission of Veloce Publishing Ltd. Throughout this book logos, m names and designations, etc, have been used for the purposes of identification, illustration and decoration. Such names are property of the trademark holder as this is not an official publication.
Readers with ideas for books about animals, or animal-related topics, are invited to write to the editorial director of Veloce Publis the above address. British Library Cataloguing in Publication Data – A catalogue record for this book is available from the B Typesetting, design and page make-up all by Veloce Publishing Ltd on Apple Mac. Printed in India by Replika Press.

CONTENTS

DEDICATION AND TESTIMONIALS

To my lovely Mum, who gave me the greatest gift: not just her love, but her belief in me.

To my long-suffering husband, Kevin, who is my partner in life as well as in the K9 Project, and stoically puts up with all my harebrained schemes.

To all our dogs, now, in the past, and in the future, hopefully. Especially Cassie.

Cassie (Courtesy Graham Fisher)

To all those who have supported the K9 Project in so many ways: It has all been very much appreciated.

And, especially to all the wonderful people who shared their journeys with us and our dogs. Your bravery, strength, and courage are my constant inspiration.

♥   This is a truly inspirational book, and the true stories are told in such a way that you feel you could be sitting, having a cup of tea with Chris, as she tells you how these dogs shaped her life and the lives of others, sometimes initiating change, aspiration, and healing. I hope this book will help to gain more visibility for this incredibly important project.

Janey Lee Grace, author/PR/media training

♥   Oscar is a Labrador who has completely changed my life. We've formed a bond of the kind that I never knew existed, and I've learned much more about myself. I've gained confidence, which is an important part of self-love, and I've stretched myself in new ways. It's all these things and many, many more that generate a bond of the kind that I hadn't imagined before Oscar arrived in my life. I suspect that, if you have a dog or other animal in your life, you will be able to relate to this.

It is therefore my pleasure to recommend this book to you. It is all about how dogs can help people; not just their owners, but others they meet along the way. The K9 Project dogs brighten many people's lives, and help them to grow in confidence and stretch their wings.

Dr David Hamilton PhD, author and speaker

♥   Some say the country is going to the dogs. But, for Chris Kent, dogs are the answer to so many social ills. This book charts her journey, step-by-step, from the germ of an idea to a thriving social enterprise. It's thorough, touching, and perfect, if you feel muzzled and need inspiring!

Robert Ashton, author and social entrepreneur

♥   Love Woof and Wonder Publishing endeavours to support the most worthwhile dog organisations and publications, and we are thrilled to support the K9 Project. The hard work and dedication that has gone into the project is commendable: a true labour of love that shines and encourages others to shine, too. We have watched both dogs and people flourish under the guidance of Chris Kent and the K9 Project programmes.

Dogs actually have the biggest heart per body mass of any creature, and the K9 project is a wonderful way to bring the non-judgemental love they have to share to light. Long may the project continue to transform the lives of the people and dogs it touches!

Caroline Griffith, energy healer, author, and creator of the Canine Flow Training Method

♥   I'm totally hooked on the book, and so honoured that I've had the chance to have a sneak peak: it is just amazing, Chris; I've been smiling and crying all through the bits I've read so far, and laughing at some bits, too. it's just totally wonderful, and my kind of book. I will be sharing it everywhere: everyone should read it!

Maria Daines, singer/songwriter, performer, animal activist

♥   For most people, having a dog in their lives is like having another best friend with four legs, but Chris and her team at the

K9 Project realised the unique ability of dogs to be much more than simply a companion.

For those young people struggling to find their way in this world, the K9 Project has bridged the void between a future with prospects and one without. Chris has written a book that will make you laugh and cry as you discover the power of our canine buddies, and why they really are 'man's best friend'

Pen Farthing, founder of Nowzad Animal Charity (Daily Mirror Lifetime Achievement Award; CNN Hero of the Year Award; RSPCA Animal Hero Award)

♥ If you love dogs, and know that they have the power to change lives, you need this book in your life. Chris Kent takes the power of the human-animal bond to a whole new level. The beauty of this book is its raw honesty, combined with true passion for the development of both the human and the animal. Chris, founder of the K9 Project, shares stories of the people and animals who have made the project the incredible success it is.

It hasn't all been a bed of roses, and, for anyone who works in this field, it is refreshing to know that, even if it doesn't work out the way you'd hoped, there are always other success stories to focus on, and other people and animals who will benefit.

This is a beautifully-written, honest, and powerful reflection that draws you in from the first page. It's accessible, enjoyable to read, and will be of interest to dog lovers, professionals who work with animals, and those seeking a new way to engage with young (and not-so-young) people.

Don't underestimate the deliberate simplicity, though. Chris is at the top of her game: she doesn't just have years of multi-faceted experience, she has made it her mission to keep learning. Chris lives her passion: a true inspiration to anyone who understands the real potential of the human-animal bond.

Marie Yates, founder of social enterprise Canine Perspective, author and entrepreneur

♥ This is a hugely inspirational book. Chris Kent's account of how the K9 Project came about and grew (and has kept growing) is heartwarming, uplifting, heartbreaking at times, and suffused with deep compassion and her love and understanding of dogs and people. It takes a great deal of skill to write a book that makes the reader feel as if they are having a cosy, one-to-one chat with the author, and Chris has achieved this.

The personal stories of those rejected by society, and whose lives were turned around by the K9 Project, are profoundly moving. The stories of the dogs involved with the project illustrate how much we can learn from our canine friends, and how generous they are with the unconditional love, which every living soul yearns for.

Hounds who heal is an honest, thought-provoking book. In chapter 29, Chris writes 'Life is made up of many small things which have great significance for me.' This book shows how something as simple as acceptance from a dog can impact on, and ultimately help to transform, a life. The work done through the K9 Project, and the people behind it, demonstrates the extraordinary power of caring amd compassion.

Lisa Tenzin Dolma, author, and founder of the International School for Canine Psychology & Behaviour (ISCP)

Foreword

There is something special about being in the company of dogs. As playful puppies, their presence makes us smile and laugh out loud, and fills ordinary days with happiness. As they grow old with us, their personalities shine, and they become our loyal, loving companions to the very end.

Animals have the capacity to influence positive emotional responses. They can lift our mood, improve our health, and make us happy. These bonds are not hampered by language or confined by borders. There is an emotional connection which transcends difference.

Recognising that everyone in life needs to feel connected, Chris Kent founded the K9 Project to help vulnerable and isolated individuals. Rather than working solely with well-behaved dogs, Chris valued all dogs for themselves, seeing their potential to make a difference.

Chris describes herself as 'an ordinary person who has managed to create something extraordinary' with a 'motley crew of mixed breed ex-shelter dogs ... and some very special people.' In facilitating so many life-changing canine companionships, Chris has proven to be an extraordinary pioneer in canine-assisted therapy. Her altruistic work deserves wide recognition and praise.

Hounds who heal is an honest and humbling account of life-changing unions between humans and dogs, and a beautiful example of how the bonds forged through compassion and shared experiences can help bring about new purpose, perspective and direction to our lives.

An inspirational book; one which you will feel privileged to read.

Dr Daniel Allen, Animal Geographer,
and founder of Pet Nation

Preface

Jamie

I wish I still had the 'before' photograph, as it shows everything so much clearer than I can explain. I can still see it clearly in my mind's eye: but I guess that doesn't help you much!

The first time we met Jamie he didn't speak at all. We were running a programme called K9 Confidence at his school, and he came into the room, and stayed for two hours without saying anything at all. Not to us, not to his fellow students; not to the dogs.

The second time he came we took the first photograph. In it, he is standing, shoulders hunched up to his ears, eyes focused on the ceiling, arm stretched out stiffly, and Izzy the dog at the end of the lead pulling away from him, trying to get to some other students to play. Jamie isn't interacting with Izzy, or his classmates. He looks deeply uncomfortable, off somewhere else in his head, stiff, silent.

In the second photograph, the 'after' one that you can see here, he is sitting with his head held high, shoulders down and relaxed, eyes open and warm, and grinning as the rest of the group applaud him, I can't remember for what. It didn't matter really, because by then, his relaxed attitude wasn't unusual. He had changed so much in the past twelve weeks that teachers and youth workers used to ask 'What have you done to Jamie?'

Somewhere in the middle of all that was a bit of Ruby the Staffy eye contact, the resultant oxytocin production facilitating a surge of happiness and feeling of connection.

Science, not magic – right?

There was also a sprinkling of a feeling of power and confidence that comes from walking a very large, powerful dog called Taz, who, deep inside, you're actually pretty scared of. More science.

Add to that mix the extra ingredients of beginning to feel you belong in a group; becoming the project's official photographer, singing on a charity CD, manning a stall, fund raising for charity,

Ruby and Jamie
(Courtesy Kevin Willats)

writing song lyrics, and standing up in class telling jokes. Stir, blend, keep the dogs involved ... it all adds up to powerful science and powerful results. And maybe a sprinkling of magic.

Jamie once said to us, "I blame you, really."

"What for?"

"Well, before I knew you I didn't do anything. I didn't go anywhere, didn't speak to anyone; didn't know anyone. I just didn't do anything. Now look at me – I'm writing song lyrics for other people, being a stand-up comedian in class, taking photographs, going out, making friends. It's really all your fault!"

I think I can live with that kind of blame.

Dogs and young people – a magical combination indeed.

1 ♥
DOGS SHAPING LIVES

 I hope you like dogs. This book is a celebration of dogs, so it will help if you do. It's also a celebration of the human spirit, so I hope you like people, too. I understand that sometimes relationships with animals seem easier than those with people. But, for me, the two always go together. The extra effort you sometimes have to make with people is usually worth it, but you need to be prepared to look at yourself pretty deeply, since people, in my experience, are not always as forgiving as dogs.

So much has been written about dogs: books and articles about how we should care for, feed and train them, how to test their IQ, play games, teach them to succeed in the show ring or at a doggy sport or discipline. So many different training methods, ideas, and opinions which can seem, to me, at least, sometimes contradictory and overly-complicated.

One thing we know for sure is that dogs have been our constant companions for thousands of years. Today, we know dog owners recover quicker from heart attacks than those who don't own dogs. We know that stroking a dog reduces stress levels; that eye contact with dogs produces the feel-good chemical oxytocin, and that eye contact with us produces that wonderful feeling of connection for the dog, too. We know that dogs work alongside us in a wide variety of settings, helping both in work and leisure, increasingly leading us towards physical and emotional wellbeing – think emotional support dogs, guide dogs, hearing dogs, seizure alert dogs, therapy dogs. Dogs as heroes, therapists, and healers.

This book is about my motley crew of mixed breed, ex-shelter dogs. They are not particularly well trained; there's nothing exceptional about them. They can all do a bit of something, but they haven't had the chance to excel at any particular doggy activity. They're definitely not heroes. But they're an entertaining bunch, nevertheless.

This book is also a bit about me (I was told it had to be, really). How I came to be where I am now, what and who inspired me to do what I do. Some of the highs and lows; joys and sorrows; laughter and tears. I'm definitely not a hero, either, or a doggy angel rescuer, or someone who has battled huge adversity to come shining through. I'm just an ordinary person who has managed to create something extraordinary, with the help of the aforementioned dogs and some very special people.

Mostly, this book is about the brave and strong people my dogs have met, and sometimes helped, and – sadly – sometimes not. Because there isn't necessarily a happy ending for everyone. But if there *are* any heroes to be found, you'll find them here. I'm sure about that. What I'm not so sure about is if anyone was 'rescued.'

I'm going to tell you about how my dogs have been instrumental in helping me set up a unique enterprise – the K9 Project. As far as I'm aware, no one else does this stuff in quite the way we do. This project uses a dog's best canine skills and qualities, which enables us to reach vulnerable, lonely, isolated children, teenagers, and adults. And once we've reached them through the dogs, the dogs then help us to work alongside them, while they shape and make changes to their lives.

Although not heroes, a dog can certainly make a difference.

But maybe you should be the judge ...

2♥

BEGINNINGS

 I had been a good primary school pupil – top of my class, or close to it. A good swimmer, I loved sports, wrote plays, and occasionally won art and story-writing competitions.

Somewhere during the transition between that cosy, creative, friendly little primary school to the extremely large, impersonal and, frankly, scary grammar school, I got a little lost. Whilst I was trying to find myself I badly dislocated my elbow, damaging nerves in the process. A period of wearing a large and heavy plaster followed – three months felt like forever. This was just before my 13th birthday, and the onslaught of adolescence. I developed a mindset that felt justified in believing that if I could not participate in swimming, tennis and other sporting activities, then it was totally unfair to expect me to be able to do maths, or languages, or history. Or anything academic, really. I mean, it was just so unfair!

So when I was getting ready to leave school, my lousy exam results were reflected in the success of my careers interview. To say it did not go well is probably an understatement. I was a typically difficult teenager by then; my approach to studying had not served me well. I had crashed to the bottom section of the class, could not understand the relevance of logarithms, or Latin, or peach farming in Australia, and had truly lost faith and interest in education. Full of those high ideas you get at 16, and wanting to fly in the face of poor exam results, I was repeatedly told I had the wrong attitude.

Despite this I felt confident I was headed for something big and important: I just wasn't sure what it was yet. I knew I wanted to make the world a better place, so imagine my shock when the careers advisor said something about retail. Not really understanding what that meant, I was horrified to discover it referred to shop work. Having worked in a shop every weekend for the past three years, this wasn't exactly what I thought of as the changing the world kind of stuff I envisaged myself doing. College was out of the question, as my father was of the 'I didn't go to college so neither are you – go out and get a job' view. (Actually, that was the best thing that could have happened to me, but he didn't know that then: at least I don't think he did ...)

So, what to do? I liked people, books, and animals. In those days – 40-something years ago – the only careers with animals either involved picking up various kinds of poo, or dealing with blood and needles. I wasn't so keen on one and fainted at the sight of the other. Animals out, then! I wanted to work with children, but was still one myself. So that was childcare out for now, too.

And that's how I found myself in a library, which may sound boring but, in reality, was exactly the opposite. It was the greatest library in the world! I was surrounded by books and worked with some very interesting and eccentric people. The person in charge, Celia, was a white witch; the reference librarian was about 90, spent most of her day on the top floor with endless volumes of dusty reference books, the daily papers, and the succession of homeless people who came in to get warm; my main colleague was six feet tall, had amazing red hair, and wore extremely high heels and lavender eyeshadow. She had figured out that, as people were going to stare at her and talk anyhow, she might as well give them something to really look at! The children's librarian built castles out of desks and chairs, and sent me off to do story-reading to schools, so I even began to fulfil my desire to work with children. There were no animals, but plenty of books about them, so no poo-clearing or needles and blood to faint at, either.

It took me another 36 years to find the way to bring together all of my hopes and dreams.

This is the story of how that happened – and, once it happened, it happened very quickly.

CARLY

I guess that Carly was the dog who started it all really. Not that I knew it then, of course.

I was 25, had qualified as a probation officer, and moved to work in a new town in Hertfordshire. It was my first proper professional job, and I loved it. Apart from all the criminal court work – writing reports, supervising offenders on orders, and visiting people in prison – we also operated as Court Welfare Officers, working with divorcing families, to help them come to sensible arrangements about contact with their children. In my naivety, I really believed that separating adults would always have their children's best interests at heart, and allow reasonable and pleasant contact to continue with the absent parent. It was a steep learning curve to discover that two adults who had supposedly declared undying love for each other at one time could be using their children to hurt and punish each other.

Mostly, it was the children who suffered, and, as court welfare officers, we would have to make recommendations to the court about which parent should have the children living with them, and what the contact arrangements should be for the other. The least popular part of the role was usually supervising contact between children and the absent parent: a difficult and often painful experience for all concerned, which usually happened at a weekend!

So, there I was at 9.30 one Sunday morning, at Tom's house to collect him for a meeting with his Dad, which I had to supervise.

Tom ran out to greet me, saying "Come and see what I've got in my wardrobe." To my surprise, the pretty family dog had given birth to a litter of pups, which were nested deep in the bottom of Tom's wardrobe, curled up on his school clothes.

I heard a voice say "If one of them is a little girl, can I have one?" and realised that the voice was mine!

Well, one of the litter *was* a girl, and have her I did: Princess Carly came into my life and brightened it for the next 18 years.

It was a miracle she lived that long because I knew nothing about dogs, really, and I must have made every mistake in the dog book. I was one of those ignorant people who thought having a dog could not possibly be hard or difficult. I mean, everyone had one! Despite my failings as an educated owner, however, I loved her dearly, and learnt as I went along. Somehow, we made it through.

Once Carly came to live with me, a very practical problem quickly became evident: I worked full-time and lived in a third floor flat (see, I told you I knew nothing!) Fortunately, back then, working arrangements were a lot more flexible; the term 'risk assessment' had not been thought of, and my boss and colleagues were entranced at the thought of having Carly join us in the probation offices. Our office was friendly, informal and buzzing, and situated at one end of the Courthouse, with the Magistrates' Court in the middle, and the caretaker's flat at the other end. Our resident caretaker, John, was pretty relaxed: happy to hoover up dog hairs, and didn't complain!

This is how and where my appreciation of how dogs can help in all kinds of situations began. It was 1979 and pets as therapy was not the common practice it is now. I had no understanding of the science behind it all: Carly was at work with me because I wanted her there; not because I thought it would help my relationships with my clients.

How wrong could I be?

And so, Carly became the probation office dog. She was a joy to have around, and soon demonstrated her 'added value' by getting people to relax and talk. Despite her diminutive size and dainty prettiness, she was actually quite hardy and brave, and loved most people. A lot of the younger lads would pretend they didn't like her as she wasn't tough ... but they still loved a cuddle. Shy, embarrassed, hesitant people talked to her rather than me, and we often ended up talking *about* her, or discussing their dogs.

Carly broke the ice. Stroking her would calm my clients, and quieten their nerves. Being around in the room, meant she reduced any intensity of sessions in a positive way, and defused any potential aggression. I never had any trouble while she was in the room.

Her choice of people she liked was interesting and revealing. She would sometimes wander out of my office, and go jump on someone's lap as they sat waiting for their appointment. Her favourite was a chap who had robbed a bank – she loved him! I later discovered he had been the driver, conned into doing the job. He was the only one caught, and had served prison time. I always thought he must've been a nice bloke if Carly liked him.

An example of the opposite reaction occurred one day when Carly was being cared for by a student who was due to interview a sex offender. When he walked into the room, apparently, Carly went crazy, and would not stop barking. She had to be removed from the room in the end.

She also developed a liking for the smell of alcohol; so much so that her entire body language changed and she went a bit – well – woozy is the only way I can describe it. Quietly woozy. This was pretty useful, actually, because when people who'd been drinking came to an appointment, she would climb onto their lap and go 'quietly woozy.' I could then ask "Been drinking?" Although they might deny it, they knew I knew, but didn't know *how* I knew. I gained the reputation of being something of a witch as a result!

Carly died in my arms at the age of 18 one Easter Monday. She had endured a major operation at 13, outlived Susie and Oscar – much younger dogs – and survived and thrived in a mixed group of lively and large dogs. She'd been my constant companion through house moves, relationship break-ups, and family bereavements. I was bereft; I miss her every day.

I learnt lots of lessons from Carly. Here are two –

♥ Small and fragile-looking, she had amazing strength and perseverance: appearances can be deceptive.
♥ It doesn't always matter what we talk about – as long as we are connecting there is hope for a relationship and a way forward. Keep trying to find a way to connect.

It took a long time to work out exactly what very important seeds she had sown, and how they would flourish and grow, once I allowed them to.

I think she would have been a fabulous dog to have on the K9 Project.

But then, perhaps she is here, anyhow.

4♥
Dogs, dogs, and more dogs

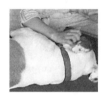

Somewhere along the line – way before setting up the K9 Project – I left the probation service, moved area, and went to work for the Youth Justice Team, where I set up a family support team for adolescents at risk of going into care.

By that time we knew that taking teenagers into care did not really help them or their families a great deal (too much, too late!), and, often, with some support, families could be helped to stay together. This could help keep everyone safe, reduce the conflict that often occurs in teenage years, help parents understand what is happening, give the young people support and guidance, and sometimes achieve a good outcome.

During this time I met Kevin, and we moved in together. At the time I had my lovely Carly, and a new dog called Susie – a little Collie-type, a nondescript black dog, who had been badly neglected – from the RSPCA. Black dogs can really get stuck in the rescue system, and are less adoptable: rescues call it 'black dog syndrome.' Susie had a bark like a seal, and an endearing habit of bouncing when it was dinner time: whilst in the kennels, she had split open the top of her head, and always bore a scar there as testament to her bouncing habit. Once home the bouncing continued – it no longer hurt her but did occasionally result in a full bowl of food or water hitting the floor – much to her delight. The cry of 'Oh Susie,' was a common one at dinner time! Susie died far too soon from heart disease (she only made it to eight or so), but she had a great time enjoying every moment of her life while she was with us.

Kevin had Oscar at the time, a huge, gorgeous Red Setter he had rehomed from a couple who were moving from a small house to an even smaller flat, and no longer had room for him.

He was gentle, goofy and soft, and we congratulated ourselves on having the most sensible Red Setter ever: he did not display any of the daft and scatty behaviour we were expecting. However, we didn't realise how badly overweight he was and, as he got fitter, leaner and faster, his scattiness, naughtiness, and sense of humour began to emerge.

Kevin worked at the family farmhouse, which was bordered by a country park. Oscar began to recognise the sounds of children playing, and, as he simply loved kids, would take himself off to the park and play. Calling him to 'stay' would elicit a glance over his shoulder, a flick of his lovely long red ears in your direction, and his swift disappearance. Sometimes, we received a phone call telling us that a family had rescued him, and taken him home with them, and Kevin would have to go and collect him; usually to find him regally installed on their front lawn, complete with paddling pool, biscuits, and happy children playing.

His adventures took a serious turn, however, when he escaped when Kevin wasn't there and he was picked up by the police on suspicion of attacking local sheep. We were pretty sure it was not Oscar, and after they made him sick and analysed his vomit we were proved correct. But it left us scared for him, and so he was no longer able to enjoy the freedom he was used to. When Oscar died he left a huge gap in our lives, literally and physically, so it was perhaps no surprise that when we went to the shelter we came back with two dogs.

Bam Bam was a Border Collie-cross-Sheltie-type, who had previously lived in South London, and had produced a litter of pups. Tiny and very, very thin, the kennel staff said that she would not come out of her kennel to see us, so we sat on the floor outside and she threw herself out of the kennel straight onto Kevin's lap. Decision made, then! Bam Bam came home with us and became Bonnie: a little bit cross-eyed, but with the sweetest disposition, and a few strange quirks. She especially liked removing a little Father Christmas decoration we kept in a plant pot in the lounge. She never damaged it, just removed it, leaving a trail of earth across the carpet, and showing us her toothy smile when we came home.

At the kennels the same day was Jason, a largish, black-and-tan boy who also came home with us, renamed Jake, and Kevin's dog-of-a-lifetime. Jake was a super dog, a real gentleman, master of the frisbee (how he loved to leap high into the air, giving a yip of satisfaction when he caught the frisbee in mid-air), chaser of cyclists (his one weakness), and the lead dog in our group (although we never realised that until he was gone, because he did it so quietly). After he died, confusion about who was in charge hit the rest of the dogs: we had always thought it was us; how wrong we were! Jake died far too young, too. If we knew then what we know now, I think he could have lived much longer, but you rely on the advice of vets, and, in his case, we might not have had the best. Bonnie lived to be 18, kissing me on the nose as she left us. Both are sorely missed.

We also shared our life with Tanga, a dog who had experienced the ravages of parvovirus, along with her pups. She, and they, just about made it, but she was very thin. Tanga was one of those dogs who left the shelter and was settled in at home with us in about five minutes! She loved beer, playing in water (although she couldn't actually swim), and running, and could produce an impressive whine if she didn't get enough of those things. Sadly, Tanga was also very accident-prone, and did herself all kinds of damage, costing a fortune at the vet. After she damaged her cruciate ligament and was in plaster for three months, she never really returned to full health. She made it to about 15 years of age, but was tired by the time she went, and was ready to go, I think, although we were not ready to lose her. Are we ever?

Kimmie the Collie was another black dog who we had to go and collect on Cup Final day, as I was certain that if we didn't she would be gone. Kevin is very patient, and it wasn't a Tottenham Hotspur final, so he happily left the sofa and television set and came with me to take a look. Despite being a 'failed' sheepdog, Kimmie was the best trained dog we have ever had. She was my holiday dog, coming away with me on horseriding trips and camps. We lost her when she was 16.

Kevin has not been a good influence as far as acquiring dogs is concerned. If I say I think we need another dog, he just says 'yes, okay, then.' When a friend, Debbie, told us about a Mastiff-cross-Shepherd pup who had been born at the stableyard where she kept her horses, and who couldn't be sold as he had been trodden on by a horse and had a huge scar, we thought we should take a look. I mean, why wouldn't you?

I went to the yard with Debbie, and by the time Kevin got there this lovely pup was sitting on my feet. As Kevin drove by I pointed at the pup, and he gave me the thumbs up. That was it: Beau was in! There's no such thing as a free lunch, of course, and, although he was free to us, he quickly incurred a large vet's bill when we realised his scar was not good, and examination revealed it was infected, which required an operation to remedy.

Beau was a big, beautiful, gentle giant who truly brightened our lives. He died at 12: a good age for such a large dog, but the suddenness of it was a huge shock for us all. He went to bed well one evening; got up the next morning and collapsed, and had died before the day was over. It was the first time we had not watched a gradual decline occur in one of our dogs, and the shock was debilitating for a while.

Some people think that if you have multiple dogs you cannot love them as much as if there's just one, and so their death is easier to bear. I have friends who have only had one dog, who say they could not go through the pain again, and so would never have another. For me, each doggy dying has a cumulative effect – you relive each loss again and again, and each seems to get harder. But I am willing to bear the pain of loss for the joy of having a dog in my life, and hope there is never a time when I feel I cannot share my life with one ... or two! I believe giving a loving home to another of the many homeless dogs there are is the best way you can honour your dog's memory. And give yourself some joy into the bargain.

Visit Hubble and Hattie on the web: www.hubbleandhattie.com
hubbleandhattie.blogspot.co.uk
• Details of all books • Special offers • Newsletter • New book news

5♥
MILES

 I also blame Miles for the project, too.

I wanted to do horse therapy – equine-facilitated learning – and had done lots of training, working my own little herd in natural horsemanship, improving my skills; getting to understand their intuitive processes. I had even done a staff training session that had been a resounding success as far as team-building was concerned. The feedback was fantastic; the pictures ace. I was inspired!

However, the logistics of moving horses and hiring venues were a bit daunting, as was my horse's sensitivity as far as working with others went. I was beginning to think I might not find a way to make it happen.

On a sunny Sunday morning, Miles sat at our large kitchen table with my husband, Kevin (at some point in the process we decided to get married), and I. We had befriended Miles, and were giving him some time away from the children's home where he lived. He was a large lad with substantial learning difficulties, and had been in care for much of his life. His was a story of neglect, abuse, deprivation, and poor parenting, as his mother battled her own addictions. There was no dad about, and Miles had a complex medley of syndromes all his own.

At 15, Miles wanted out of the care system and back home, however inappropriate the professionals thought that might be. The staff at the children's home where I provided training and consultation were struggling to help Miles control his temper, and become involved in positive activities. His behaviour was escalating (or descending) into aggression and assault, as he struggled to make sense of who he was and what the future held for him. He was throwing his weight around, literally, and the staff were beginning to find him intimidating.

Their frustration, and his, was increasing as different ways to control, motivate and encourage him were tried. They all really liked him, and knew that underneath all the bluff, bluster and threats, there was a scared, lonely, confused little boy. But absolutely nothing was working.

I'd been asked if I could provide Miles with some downtime at weekends. He liked dogs – and we had plenty of those – so he began weekend dog walks with us, and was doing well. We decided, however, that it couldn't all be fun and laughter: Miles had to earn his dog-walking and play time. We devised some written exercises to keep him occupied, and he began putting pen to paper (for the first time in months) in exchange for the fun.

On this particular day, he watched, stunned, as our newest dog, Ruby, a Staffie-cross who had been saved from death at the pound after her seven days were up (she was a little powerhouse, with very little in the way of training) came barrelling into the kitchen.

Ruby jumped on Kimmie, the elderly Collie, knocking her to the floor, then ran into Cassie, and asked her to play by throwing herself at her, then play bowed to Buck the Doberman who made himself scarce. Ruby dived on her bed and shook it, all the while furiously wagging her tail. She stopped, looked around, and dived towards Taz, hitting her head on the table leg as she ran past, playfully pouncing on his chest while he was trying to snooze in the sun.

"What's wrong with her?" Miles asked.
"Well, nothing." I replied.
"What's she doing? She's being a bully."
"She's playing – kind of."
"But none of the others want to play with her; she's being really rough! Poor Kimmie!"
"She has a lot of energy. She doesn't always know what to do with it. Sometimes she just gets too excited to stop." I explained.
"Yes, but it's not fair on the others. She's being a bully. Can't she tell they don't like it?"
"No – why do you think that is?"
"Dunno. Is it because she doesn't know how to?"
"Maybe. The other dogs know how strong she is and don't always want to play."
"Is it because she's so excited she can't stop? I get a bit like that sometimes."

"How come?" I asked.
"Well, I play fight."
"Do the other kids like it?"
"No."
"Do they ask you to stop?"
"Yes."
"And do you?"
"No."
"Why not?"
"Dunno. S'pose I get to a point and I can't stop. I kind of lose control. Then the staff come along and tell me to stop and I hit someone."

Pause.

"So, what can you learn from watching Ruby?"
"She needs to find something else to do when she feels like that. Can I help her find a different way?"
"Maybe. If you want to." I said.
"Yes, I do. Maybe that will help me stop 'doing a Ruby' myself."
"Well, we can all help with that."
"It'd be good to help her and me at the same time ... I reckon we've got a lot in common!"

Miles returned to the children's home that day, keen to tell the staff that he was going to try and not 'do a Ruby' in future. Exhausted (we were usually exhausted after a day with Miles and his boundless energy, grasshopper mind, and constant questions) we sat with a coffee, reflecting on the session.

At last, Kevin said to me, "You know that thing with the horses you wanted to do ...? Well, maybe it would work with the dogs instead? It would be a bit different, but it would use the same principles, and it could also be easier! Dogs are more mobile, so they're easier to take out to people. We could do some animal welfare and responsible dog ownership sessions. We could work with students who were struggling; we could write some educational material they would understand, and we could use the dogs to help them learn about themselves: energy, communication skills, self-management _ all that stuff."

And you know what?

We could ... and we did!

6♥
LET'S GO!

"If you think you can, or you think you can't, you're probably right!" Henry Ford

So that was it, then. Decision made. How hard could it be to get going? We knew what we were going to do. We had the dogs – what else did we need? Mind you, there were bound to be lots of health and safety questions. And insurance ... that might be challenging, but, a name, a logo, branding, and off we go.

Erm, not quite ....

The name was simple. We had already been calling it the 'dog project' or the 'K9 project,' so that's what we settled on – The K9 Project. A friend recommended his friend's graphic design company, which produced a smart, professional-looking logo, headed paper, business cards, and all the trimmings. Done!

Perhaps a second opinion about our dogs' suitability might help. I quite fancied a bit of dog training anyhow, so we asked a friend, who recommended Charlie Clarricoates, the proprietor of a local dog training centre. She said he was great, but very direct, and very blunt. That did not put me off. I rang him, and the conversation went along the lines of –

"Hi. My name's Chris, and I have a whole bunch of rescue dogs, and we're going to run a project working with disadvantaged and challenging young people, and we need someone with credentials to assess the dogs for us. Oh, and also I would like to come to some dog training as a couple of my dogs could do with some. Oh, and I don't really want the dogs to be too perfectly behaved as they need to have a few issues to mirror those of the students."

I paused for breath. There was a long silence on the other

end of the phone while Charlie no doubt wrestled with the desire to put the phone down on this mad woman.

"Perhaps you'd better come over," he said. I've no idea why he didn't just say, "Sorry, but I'm too busy/on holiday/washing my hair" or just about anything apart from what he did say! We met up, I explained it all slightly more coherently, and Charlie, I think, was intrigued to find someone who actually wanted her dogs to have a few problems or issues. Usually, people want him to help turn out perfectly-behaved dogs. As a people person as well as a dog person, he was also intrigued by our idea, and 'got' what we were trying to do: he very much uses the model of working with the owners to help the dogs, so understood much of the metaphorical learning we were aiming for. He also likes to help people, so understood our motivation.

Anyhow, he agreed to assess our dogs as long as I went to some training classes, and this was the beginning of a long term relationship between Charlie, his wife Jo, his training centre, and the K9 Project. We did some dog training there, and went on to get a range of dog behaviour and training qualifications through an educational establishment the two of them were involved with at the time.

Charlie let us use his facilities: an absolute stroke of luck as they were the best training facilities in the area, which meant we could offer a lot of activities we would not otherwise have been able to. There were large fenced paddocks, recall lanes, an indoor classroom/lecture theatre, toilets, and poly tunnels with heating and lights that were great in winter!

We ran (and still do run) regular sessions there with students, as well as summer scheme activities and volunteer training. The young people love to watch dogs swimming in the centre's hydrotherapy pool. If we are really lucky, and Charlie has time, he can sometimes be persuaded to come along and run a session. His way with the young people is fabulous – they respect him entirely, and, of course, his dogs are far better behaved than ours!

So that was our name, logo, dog assessor and first supporter arranged. What about insurance? Well, that was relatively easy. A chance meeting at Crufts yielded a happy conversation and a reasonably-priced insurance policy. Legal status? Originally, the project was set up as a partnership to see if it might take off. When it seemed it would, we realised we could not attract grant funding as a partnership, so needed to create a non-profit organisation. The thought that we might ever make anything beyond a few quid in salaries was laughable then (and even more so now, but more about that later!). We sought some good business advice and set up as a company limited by guarantee, with no shareholders – and any profits had to go back in, which was our plan, anyway! This meant we could apply for some funding grants, as well as trade with anyone who wanted to pay us! The other advantage was that we could retain some control over the business, and be paid. We paid £100 to have all of the paperwork prepared; found ourselves four other directors, and we were off: proud directors of a limited company called The K9 Project (Training and Development) Ltd.

Simple? Well, that bit might have been ...

VISIT HUBBLE AND HATTIE ON THE WEB: WWW.HUBBLEANDHATTIE.COM
HUBBLEANDHATTIE.BLOGSPOT.CO.UK
· DETAILS OF ALL BOOKS · SPECIAL OFFERS · NEWSLETTER · NEW BOOK NEWS

27

7♥
FIRST ASSIGNMENTS

'She took a leap of faith and found her wings on the way down'

I saw this quote on Facebook, and it struck a chord, reminding me of how it felt to learn as you go, and how little I knew about what I was doing when I started.

We'd set up our limited company and organised our insurance. Now, where could we get some work?

The first place to give us a try was a local unit within a mainstream secondary school, which took students from all over the county. Most had Statements of Special Educational Needs (a formal document detailing a child's learning difficulties, and the help that will be given), and learning difficulties of some kind, either diagnosed or undiagnosed. Many also had family environments that were not conducive to education, or rules, or positive outcomes.

The place was wild. So many young people, aged 10-17, with so many problems, in such a small place. The staff were often young, untrained, and unprepared for the kind of problems they would encounter. One young staff member told us she felt scared every day. Waiting in the common room for the students to pour out of class was an education in itself. I learnt to plant my feet, centre myself deep into the earth, and breathe deeply, otherwise it felt almost impossible to maintain position as I was buffeted across the room by the force of the students streaming past, and in some cases into, me. The majority of the staff cared for the young people deeply but seemed unable to do anything to attempt to engage them, and the concept of boundaries or rules appeared non-existent.

For us with dogs, this simply wasn't going to work. We had

to establish strong boundaries or we would be putting our dogs in jeopardy: unthinkable. This was also our first piece of work, and we badly wanted to be successful.

We didn't work within the school but used the local park, youth centre, or dog training centre, working individually or in small groups. There was a limit to how many young people we could manage safely. Sometimes, the school sent staff along to 'help,' which was usually counterproductive as they brought the school 'atmosphere' along with them. One staff member spent most of the session asleep on the sofa – not exactly the role model the students needed. We asked that he didn't come again.

The young people were allowed to smoke with parental permission, which meant that children as young as ten were puffing away during school hours. Call me old-fashioned but there was no way I could allow this, which meant that some students would simply not come to sessions. So be it. Occasionally, some months later they might decide to accept our rules – no smoking, no phones, and no computer games – and turn up.

We established a regular pattern of students, and worked with some for two years. Some of them came from such chaotic and heartbreaking backgrounds I felt that I just wanted to save them, keep them warm and safe, and let them cuddle the dogs. A fair amount of this did happen, but, in the long term, it wasn't what K9 was intended for, and would not help beyond the immediate moment. The students needed to learn to function in groups; be able to learn, and control their behaviour. It was challenging for them, and us, as some of their stories later in the book testify.

One lad, Tommo, used to say that when he won the lottery he would buy the project, and we could work for him. He would pay us well and come to the project every day! We were the only place where he ate anything, or even *did* anything! If you asked Tommo what he'd done in other lessons, he might say 'swimming,' but what that actually meant was that he sat in the seats by the pool playing a computer game. When he said he'd been 'fishing,' this usually meant he sat in the van playing a computer game. 'Art' usually involved sitting in the art room – yes, you've guessed it – playing a computer game.

Working with this school, we learnt so much about safety, flexibility, and how working with the dogs could pan out. We formed some great long term relationships with the students, and saw huge signs of progress. Working with them twice a week for two years, 36 weeks a year, demanded a big commitment for some of the students, most of whom had the attention span of a gnat (unless it was for a computer game!) It was a joy to watch them grow and develop whilst working with the dogs. It was also challenging, heartbreaking, exhausting, frustrating, fun, and deeply emotional. Many of these young people were so convinced by the age of 12 that they were totally useless, there was no convincing them otherwise. It is difficult to motivate anyone who feels they have no future, and nothing to offer; it is equally hard to work out how to inspire someone who appears to be completely disinterested.

We also learnt a very important lesson about how those who commissioned our services viewed our involvement, which was also relevant to how they viewed the young people they educated. Two days after the end of the summer term we received a standard letter from the school to say thanks, but our services were no longer required. Bang. Just like that. I was in total shock. I later discovered the school had received an inspection visit which had not gone well, and was undergoing a complete change of regime and ethos as a result. Nothing personal.

But I kept wondering what the students would think when they returned for the new term and we weren't there. What would that say to them? Some had been let down so many times by adults; this would only reinforce their view that adults were untrustworthy, unreliable, dishonest, transient, and uncaring. It also deprived them of the educational opportunity of experiencing an appropriate goodbye, with positive feedback from us, and a chance to say goodbye to the dogs. For some of them, the dogs had become a very important part of their lives, and kept them going. To deprive them of a chance to say goodbye seemed unnecessarily cruel.

I never really knew what the motivation was for the school's approach: maybe it was trying to avoid a riot or painful experience for the students. I suspect the truth of it is that no real thought was given to how best to handle the situation.

I do know that it had an impact on some of the students, one of whom returned to school for just one day, before refusing to go again. He always said he would stop going if we were not there. I wonder whether if we'd had the chance to say goodbye to him, could we have helped him make different choices? We'll never know.

The situation left us with an uneasy feeling about just how highly young people might be valued by those entrusted to care for and educate them. Sadly, this was not the only experience we had of the system failing to support students so obviously in need of it.

INSPIRATION AND ESTABLISHING LINKS

We looked to America for inspiration and practical ideas for the K9 Project. There wasn't really anyone in England doing what we wanted to do at the time, so we Google-searched and it highlighted many schemes in America: some of them market leaders in their chosen fields. Animal Assisted Therapy was a far bigger industry at that time there than here, and many hours were spent researching websites. There were many fantastic projects with prisoners caring for and training homeless dogs, and helping them be rehomed. All really exciting, effective, and inspirational, but not so relevant for us, as we would be working in the community, and not always with dogs in shelters. I had already researched those programmes that use prisoners to work with wild Mustangs and horses off the range, so was not surprised to find shelters and kennels attached to prisons, where prisoners could take care of the dogs. The adoption statistics are fantastic, and recidivism rates for the prisoners are greatly reduced.

Then I found Glen Highland Farm in New York State, a Border Collie rescue that, for at least three weeks in the summer, opens its doors to deprived New York inner city teenagers who have never seen the countryside, or experienced anything beyond their own neighbourhoods.

This is a magical project, full of joy and laughter, located in a beautiful setting surrounded by woods and streams, and sacred Native American places. The project chose Border Collies as they are bred to want to connect to a person, so are very good at bonding with their chosen handler. The residential programme also includes lots of other activities, but the dogs are at the core, and each young person has a dog to be responsible for during the programme, to train and help them

continued page 41

Girl Power: Cassie, Izzy
and Ruby
(Author collection)

Billy
(Courtesy Sylvie Gonera)

Summer fun
(Courtesy Sylvie Gonera)

Billy, hoping for a
tummy rub
(Courtesy Sylvie Gonera)

Christian and Billy
(Author collection)

Celebration
(Author collection)

Cassie working alongside Austen
(Author collection)

Getting close
(Courtesy Sylvie Gonera)

Kieran and Billy
(Courtesy Sylvie Gonera)

Top left: Connecting (Author collection)

Top right: Taz in contemplative mood (Courtesy Kevin Willats)

Left: Kristian, Buck, and Ruby (Courtesy Sylvie Gonera)

Above: Billy likes to get up close (Courtesy Sylvie Gonera)

Below: Billy and Tyler (Author collection)

Top left: Taz can get lively, too
(Author collection)

Top right: Taz loves a cuddle as much
as anyone (courtesy Kevin Willats)

Above: Ruby catching some rays
(Author collection)

Above right: Loving a cuddle (Author collection)

Right: Ruby Hippy Chick (Author collection)

Right: Buck (Courtesy Sylvie Gonera)
Above: A reassuring touch (Author collection)

Left: Cassie helping with guitar practice ... (Author collection)

... and with difficult conversations (Author collection)

Doing the 'people jump'
(Author collection)

Cassie jumping over Jack
(Courtesy Grant Auton)

Left: A problem shared (Author collection)

Ruby connectin
(Courtesy Sylvie Gone

be adopted. This was more like it, but still hard to replicate here in the UK. I found myself wondering if I was going to have to start my own shelter!

Next, I discovered K9 Connection, based in Los Angeles, which offered 15-day programmes for high school students who had lost their way, and were not achieving as well as they might. These students were paired with shelter dogs in need of homes; they would train them every day, and complete classroom-based activities, too. This felt closest to what we were hoping to achieve, and what could be replicated here. Katherine Beattie, one of the founders and directors of K9 Connection, was so helpful and encouraging, freely sharing lots of ideas and information. This organisation remains my main inspiration.

Surprisingly, Cesar Milan made me realise how sensitive dogs can be when he demonstrated how changing an owner's energy can result in something different in a dog. At that point, I don't think I had appreciated this, as I was so used to working that side of things with horses.

The metaphor of dogs in shelters needing homes and a second chance is a powerful one, which I wanted to replicate. At that time, the shelters in our area were not interested in working in partnership, which was understandable, really, as they were busy enough trying to save canine lives, and had no time to worry about people, too! We could use our dogs – mostly ex-shelter/homeless dogs, but now safe in their forever home – as they would still provide a visual learning experience, and would be helpful for the therapeutic side, as well as teaching about dogs, dog training, and all the other life skills. But the experience would not be nearly as powerful as helping a dog in need find a home.

As a firm believer in adopting from a shelter rather than buying from a breeder, I wanted to make the most of this point, which was particularly important as we would be working with future generations of potential dog owners, who might not have had the best role models of responsible dog ownership thus far in their lives.

If we were not able to regularly work with rescue dogs, we could maybe fundraise for them, which would reinforce our education sessions about responsible dog ownership, and highlight the continuing crisis in rescue shelters in the UK, where so many healthy dogs were being euthanised (statistics say one every hour, although I am sure it is more). In addition, it would teach new skills, build confidence and self-esteem, develop a sense of empathy, responsibility, and citizenship, and allow for creativity in developing projects.

I looked around for different types of rescue centres, and found opportunities to work alongside some of the larger ones, but also formed partnerships with smaller, independent rescues which were usually strapped for cash, and took in difficult dogs that were not as easy to rehome. I also looked for organisations that helped people as well as dogs, and found Canine Partners, which trains assistance dogs, and also established good links with Brian and Beasley, a human/dog pairing close by who gave talks and demos.

Then I read *One Dog at a Time* by Pen Farthing, founder of NOWZAD Dogs. The book moved me greatly, and the charity's mission to help dogs make it back to live with the soldiers they had befriended in war-torn Afghanistan fulfilled the criteria I wanted. A discussion with Pen resulted in his kind offer to give an impromptu talk to a group of teenagers who we took to Crufts one year, which really inspired them. NOWZAD has since gone from strength-to-strength, and is now a global charity with a great team, but, initially, it was Pen's determination and drive that was its inspiration.

We now had two charities to fundraise for, which fulfilled the criteria of helping dogs *and* people. Perhaps more importantly, though, both charities encouraged young people to think and learn beyond their usual boundaries; to consider disability and war, personal strength and motivation, and the difference dogs can make in enabling people to lead fulfilling lives.

We also formed a good relationship with our local branch of the RSPCA, and became accredited volunteers, spending many a happy hour walking its various residents. We always came home feeling better than when we went out, despite the sad stories of many of the dogs: their ability to find joy in a walk around a wet and muddy field always lifted our spirits. With a change in the management team, we were able to involve some of our young people in programmes there, and allowed to run dog training alongside the RSPCA staff with our young people's group work programmes.

This was closer to what I had always dreamed of ... working with the shelter dogs, helping them be adopted.

Powerful stuff, indeed.

♥9
How does this dog stuff work, then ...?

This is a question we are often asked, and, to be honest, it is not easy to give a brief answer. Funding bid applications usually allow a maximum of 200 words, and business coaching suggests the answer should be no longer than the time it takes to descend three floors in an elevator. Perhaps not surprisingly we haven't yet perfected the elevator answer, although I am going to try and do it in a chapter.

At the most basic and fundamental level, we teach young people about dogs and how to be responsible dog owners: they are the dog owners of the future, after all. We teach them what dogs need to be happy, healthy, fulfilled. How much it costs; what about insurance?; what are the laws about you and your dog? We teach them how to stay safe around dogs they don't know, and, indeed, dogs they do know. Then there's equipment, nutrition, knowing signs of illness. We can incorporate maths into calculating costs for different types of dog foods, and use IT to research breeding or puppy farming. This is the key to learning – engaging young people in a fun, informal way about a subject close to their hearts.

Alongside this we will teach some dog training: not just how to train a dog, although that can be helpful, but more about developing communication skills – how to be a clear verbal communicator, master the use of body posture, and other non-verbal signals. We will teach an understanding of dog body language first, then relate it to our own. A lot of our students have poor hand/eye coordination; even connecting a lead to a collar is difficult for many of them. They also find following the most basic instructions very hard, and usually all talk at once. A major task for us is to reduce the noise, increase attention span, slow the pace, and group tasks in small, achievable chunks to maximise success. We follow some simple steps (not always totally effective!), and try to mirror what we are asking them to do with the dogs in how we work with *them*. In essence, the fewer words – clearly articulated – the better, with accompanying hand signals or body movements; small steps, and consistency of approach, with a reward at successful completion.

A major part of what we teach is based around understanding energy: where it comes from; what can be done with it; how to control it; how to direct it appropriately, and the difference between being passive, aggressive, and assertive. What works. Through all of this, the dogs provide most of the feedback. Our role is to ask questions that promote thoughtfulness, understanding, and a questioning attitude – Did that work well? Was there anything that you could have done differently? How do you think the dog is feeling? How did that feel? Do you realise how well you did that? All the time giving lots of praise, positive reinforcement, and encouragement.

Sometimes, really powerful moments come from the students being the teacher. They cannot always cope with the dogs not doing what they want, even if their signals and cues are inconsistent, and even if the dog does not really understand what he should do. Our dogs are definitely not robots, and across our group we use different approaches to training, so students have to learn how to adapt their communications. Their frustration at the dog not doing what he is 'told' is tangible, and often leads to discussion about how it feels for the students' teachers who are trying to teach them, as well as debate about how to adapt cues, learn patience, and simply wait! None of these things is usually easy for them, and often a lot of support is needed. However, when they *do* finally get a result they're very proud, and their smiles let you know *they* know when they've 'got' it.

We also work on soft skill development (a combination of interpersonal people skills, social skills, and communication skills), and increasing emotional intelligence (the ability to identify and manage personal emotions, and the emotions of others). The development of empathy is not always easy for youngsters who may not have had their own needs met, and whose adolescent brain has not yet developed enough to understand this aspect of emotional literacy. However, many have warm hearts and feel more empathy for animals than for other people. Accepting responsibility for our own actions and behaviour may begin with being responsible for the dog on the other end of the lead, and develop into project managing a fundraising activity. Managing our own behaviour, especially temper and aggressive outbursts, is not an easy task but, again, it appears to be easier to do for the sake of the dogs, than for ourselves or other people.

Risk assessment, and thinking about what we say and do *before* saying or doing it, are also very difficult tasks for the teenage brain. Walking a dog on a lead is a good opportunity to promote this area of development. Approaching roads, or other walkers, rivers or streams, provides further opportunities for forward observation and planning. Helping each other, working as a team, being kind to the dogs, and having fun, receiving praise and encouragement, and observing non-judgemental feedback from the dogs can be a new and exciting experience for students. We work with some of the most challenging pupils who don't often demonstrate behaviour within school that is worthy of praise, or rewarded by staff there.

For many students with difficulties, the large size of a secondary school system, and the expectations therein, does not serve them well. Working with us in a smaller group, in an informal setting, and 'doing' rather than sitting, getting feedback from the dogs, allows them to shine, and be the best they can. We try to give them as many positive choices as possible, so that they 'own' the programme, working on mutual respect, and following the dogs' example of always being really pleased to see them. In this way they can be appreciated for the wonderful young person that they are, behind their sometimes anti-social behaviour; rewarded for their strengths, and given reasons to feel good about themselves.

Another key aspect is the physical presence of the dogs – their calming influence results in a lowering of blood pressure and a reduction in stress hormones – they have huge therapeutic value. The power of physical contact cannot be underestimated, and this is especially apparent in secure settings where rules and boundaries must be heavily enforced. On one visit, a young woman had to be almost forcibly removed when she clung to Ruby as if her life depended on it. There is lots more about the science behind this in another chapter.

Talking about issues relating to dogs can lead to open discussions about a whole range of topics. The Dangerous Dogs Act and Breed Specific Legislation are great ways to introduce a discussion about stereotypes, discrimination, and fairness: all very important and relevant topics for teenagers, especially those who have experience of a 'system' – be that the care, justice, or custodial system. Killing dogs that no one wants, over-breeding, greyhound racing, and animal cruelty are all excellent issues with which to introduce debate about morality, money, animals in sport or entertainment, ethics, personal values, punishment, and deterrents. The young people are able to discuss matters without seeming too personal; gain insight into different points of view, and stimulate those parts of their brain that need developing. We often do a debating exercise where the students have to adopt a standpoint that they do not necessarily agree with, and then present it to the others. They find this very challenging, but produce some great work. All the young people we work with have difficulties of some sort, either emotional, behavioural; a developmental delay or a learning difficulty. They may be young carers, may have been bullied, or be a bully. They may have extremely low confidence and self-esteem. Often, but not always, their home experience will not have been positive. Fragmented families, neglectful environments, poor communities, a lack of boundaries or consistent support, many moves of areas and schools, poor social skills, a lack of a good role model, parents with mental health, drug and alcohol problems, domestic violence, abuse, being in the care system – all of these are common family situations.

This means that, in addition to all the education we provide, we must create a safe environment for them, and be nurturing, caring, concerned, flexible, and able to manage their often intense and fluctuating emotional moods, which can vary wildly within each session, often switching rapidly in minutes.

We do all of this whilst keeping dogs and students safe. Good behaviour management practices are essential. Using kinaesthetic learning (or tactile learning, a learning style in which learning takes place by students carrying out physical activities, rather than listening to a lecture or watching demonstrations), and lots of physical activity helps all the students, but boys, in particular. It is well researched that boys will be mostly kinaesthetic learners who need to move, experience, and do things to maximise learning opportunities. Pairing a hyper dog with a hyper student can tire out both, help them be calm, and facilitate a state of mind ready for learning.

The progress we make with each group or individual varies, depending on how often we work with them, for how long, what the mix of the group is, and what support we have from other professionals. Sometimes, the group works really well, even though on paper it should not. Other times, careful consideration of group members still results in a less successful project. We quite often have no say in who attends our sessions, and no prior knowledge of the participants. In our experience, groups work better in the morning than in the afternoon, by which time students are often already tired, have maybe had a bad experience at school that morning, and arrive already eagerly anticipating the end of the school day. Groups are best run away from school facilities. And plenty of refreshments are needed!

Working alongside the dogs is effective on many levels, and can be applied to any situation or circumstance. But the importance of having highly skilled, creative, adaptable, caring, and knowledgeable staff with the right child centred-approach and resilient nature, and the right kind of dogs to work alongside, is paramount.

VISIT HUBBLE AND HATTIE ON THE WEB: WWW.HUBBLEANDHATTIE.COM
HUBBLEANDHATTIE.BLOGSPOT.CO.UK
· DETAILS OF ALL BOOKS · SPECIAL OFFERS · NEWSLETTER · NEW BOOK NEWS

10 ♥ Connection, connection, connection!

"It is in our nature to be kind. When we don't take up the opportunities that the world presents to us each day to be kind, we betray that nature. The multiple benefits are then lost as opportunities slip away. But when we are kind, we make everyone's life just that little bit better" – Dr David Hamilton

Oxytocin is a chemical produced in our bodies; often called the feel-good chemical, the love chemical, the cuddle chemical or the chemical of connection. To be really technical, it is a neuropeptide made up of nine amino acids, produced primarily in that area of the brain known as the hypothalamus, although new studies show that it is also made in the heart. It is present in the human foetus and placenta, and stimulates contractions. When released, oxytocin travels round in the bloodstream, serving as a hormone, with several different jobs to do.

Oxytocin can be produced in large quantities when we are in love, when we feel connected to another (human or animal), when we listen to music, when we hug, when we are inspired, when we receive or bestow a warm touch; when we stroke or caress animals. Production can be two-way and affect both giver and receiver. Researchers in Japan studied dog owners playing with their pets, and found that those who experienced longer periods of eye contact with their dogs had an oxytocin increase of 20 per cent[1]. Of course, those of us who have dogs have experienced the rush of goodwill and 'lightness' when they are around, and the calming effects of stroking them: they can lift our mood in a second. So this is not really news to us.

The production of oxytocin also impacts our health: research indicates it can positively affect our digestive system, reduce inflammation and lower blood pressure, protect against hardening of the arteries, slow ageing, and assist with healing. Babies need to experience oxytocin-producing behaviour from their carers – love, touch, stimulation – in order to thrive and grow both intellectually and physically. Oxytocin helps us be more trusting, too.

One behaviour that produces oxytocin in both giver and receiver is kindness. I know this scientific stuff because I have read *Why Kindness is Good for You* by Dr David Hamilton. I have also met him, and talked about at length. In his book, he reviews much of the research about kindness, oxytocin, health and emotional wellbeing, and some fascinating insights reveal how we can produce more oxytocin in ourselves, and encourage kindness in ourselves and others. This book was a real eye-opener, as it helped me understand the science behind what we saw happening in our K9 Project sessions with young people. Additionally, it gave us guidelines about what would work best in promoting kindness and a sense of connection within our sessions, not just between the dogs and people, but also between ourselves in general, and particularly those who participate in the sessions, regardless of age.

It starts with the dogs, though, and especially those animals who can maintain eye contact (some dogs are not confident enough to maintain eye contact, or regard it as a threat), can sit still and appreciate being stroked, and who can respond in an affectionate way. If someone looks at Cassie in a certain way she will put her paws on their shoulders, face-to-face, look deep into their eyes, and then rest her head on their shoulder or chest. It is a heart-melting gesture, guaranteed to draw the 'awww' response, and make people feel good. Ruby is especially good at maintaining eye contact, gazing adoringly up into the eyes of others ... as long as they rub her tummy! (Double-whammy: eye contact + physical touch = oxytocin = connection.) One lad, who had difficulty expressing any emotion, was on the receiving end of the 'Ruby treatment,' and looked up with a glow in his eyes, saying "I think I'm in love!" Billy can find it hard to maintain eye contact, but has an endearing habit of collapsing onto his back and asking for a tummy rub, or standing on his hind legs and asking for a cuddle, which seems to work just as well. I think one of the reasons Izzy is not as popular with some students is that she can find it hard to sit still and maintain prolonged eye contact.

Once the oxytocin is flowing with the dogs, we build

in opportunities for people to be kind, and demonstrate this with our own behaviour. Kindness is contagious – especially in children and young people. They can be kind to the dogs by filling their water bowls, giving them treats, stroking them, rewarding them for good behaviour, putting them in and out of our vehicles, and looking after them in the breaks. We, in turn, are kind to the group: we're pleased to see them, we speak kindly to them, notice everything good they do, and make sure we let them know we've noticed. We are warm and encouraging, bring in treats (sweets, fruit or homemade goodies), and let them know they are important to us. If it works well we soon have a group being kind to each other, and receiving positive results in return.

Many of the young people we work with do not get many opportunities to give or receive kindness, either at school or at home. They feel disconnected from their peers at school, have often been abandoned by at least one parent, and are not exactly staff favourites within their school environment due to their behaviour: in short, they can feel totally isolated and disconnected from the world. It is hard to see young people who feel this way, but our work with dogs can at least open up other possibilities, and offer a new way for them to feel about themselves and others.

One lad, CT, had a fearsome reputation at school, because he attacked anyone and everyone: teachers, much older and bigger students, male, female, whatever. He was a very small lad, recently arrived from Eastern Europe, who had been relocated within an alien culture and didn't speak English very well. Some of what he experienced in his short life had been traumatic, sad, and scary. He must have felt like he was drowning, but he came up fighting, which only made things worse.

CT came to us as part of a group from his school, and, right from day one, showed a completely different attitude and character: he loved the dogs and gained a great deal from their physical presence. He patiently and persistently helped one boy overcome his almost phobic fear of dogs, became our kitchen helper and regularly waited on everyone in the group, making hot chocolate and toast, and washing up.

On a visit to a rescue shelter, CT was the only one who was able to cuddle a Bulldog that the other students could not bring themselves to go near. ("I'm sorry miss," said one. "I know it's wrong of me but she is so ugly I just can't!") CT replied that the dog was beautiful to him, and it was what was inside that mattered. I like to think that we got him feeling connected, firstly to our dogs, then us, the other students, and then to others, but, most importantly, to himself. I hear he is doing much better in school now.

It's possible within groups to see the contagious nature of kindness as it spreads, prompting better behaviour, improved and new friendships, empathy, mutual support, teamwork, laughter and fun, open, trusting conversations, and improved school and project work. There are times when the groups are so good, so powerful, I do not want them to end (and neither do the students, usually), I fear for them going back into a sometimes hostile environment, hoping we have had the time to have done enough to increase their resilience and improve their behaviour. Sometimes, I still feel a desire to take home a student or two (though I don't, of course!) You would have thought that, after 40-plus years in this line of work, I would be past that, but the oxytocin rush, the 'helpers' high,' the feeling of being instrumental in helping someone make positive change keeps it fresh, powerful, and meaningful, still.

With adults the same things can materialise, but it is often harder to create the situations necessary to help things happen. Mistrust may be harder to tackle, as adults have had longer to build their view of the world, may be more reluctant to reveal vulnerability, and have better defences to guard against hurt and pain. The dogs still have a huge impact, though, and the kindness model still works. We can get amazing results; they may just take a little longer to acheve. We have seen very diverse groups of people come together and bond – united by their love of dogs, and their desire to make changes in their lives.

Connecting with a dog does not replace connection with other people. I realise that, for some adults, trust is a major issue, and the dogs' non-judgemental approach helps establish that initial connection; as we often say, dogs do not care if you are fat or thin, rich or poor, if you can read or write or wear designer clothes. However, I would hope that the feeling of connection they get becomes addictive, leading to connections with people, as well.

And I guess I need to add that it doesn't always work – not for all groups, or individuals, although we don't know why this should be There was a time when I wondered whether those people who had not experienced the oxytocin rush of connection in their formative years were unable to experience it in later life: did the 'use it or lose it' rule apply here? Was it too late for our programmes to be effective in that way? A conversation with David Hamilton reassured me that it is never too late, however, because we are hardwired to produce oxytocin, and hardwired to need the things that produce it. I have seen enough evidence of it myself now to believe that this is the case, but it's good to have the science to back it up!

Of course, I get my own oxytocin rush from my dogs, from the project, from the people we meet. It's safe to say my dogs have helped me get through some hard times, brought me back from some dark places, and got me outside into the sun with the wind in my hair and a smile on my face.

And the rain on my head, obviously. This is England, after all ...

Well, maybe that's enough chat about all that kinda stuff. Now it's time to read some people and dog stories. That's really what you're here for ... right?

11♥
AND SO TO THE DOGS

 Our dogs are all different, although they are all ex-homeless/shelter animals. They all have varying roles to play, and can teach people different things. They are probably not as well trained as many therapy dogs, and we feel that this is often their strength.

People's issues are sometimes reflected in a dog's behaviour, which enables them to connect to the dog, and, by helping the dog, help themselves.

This book is not the place for a detailed analysis of how we assess our dogs; how we monitor their enjoyment of the work they do; how we constantly risk-assess for dogs and people, or how we ensure the welfare of dogs and people at every stage of this journey. But I want to reassure you that we do, constantly.

There has to be something in this process for the dogs, too: I do not believe that dogs exist purely for our convenience, and I am often concerned about the healing demands and expectations we place on them. This project works on dogs being dogs: the lessons they can teach us by being themselves, in all their glorious 'dogginess.'

We are always mindful of our dogs' welfare, and we monitor them constantly for signs of tiredness, discomfort, and distress. If they need time out – be it ten minutes or ten weeks – they get it. If there are elements of the work they do not enjoy, they are not asked to do them.

Ruby, for example, isn't keen on large groups or long car journeys; Izzy loves both. Taz likes training activities. Cassie is happy to hang out, do agility, go for long walks, swim, or cuddle. Just like the people we work alongside, we play to their strengths and individuality. There simply is no other way.

Billy

Billy, our Irish Collie-cross, resonates with people who experience severe anxiety and low confidence because he can be nervous himself. He came from Ireland, and our thinking is he was not socialised at all, or exposed to any noise; probably kept in a dark, quiet shed somewhere; possibly beaten. He was certainly bald down the backs of his legs due to laying in urine and dirt. He had such submissive body language, and his back was so curved, that the rescue centre had him checked out in case he had hip or back problems. They did not find anything physically wrong with him, but he has some very stressed-based behaviours.

If anxious he will eat – anything – in a compulsive kind of way. He has eaten a small plastic bag of elastic bands at home. I have to check classroom floors before we go in for things he might eat, or get the students to help me do so. He likes small, plastic things. Loud noises are his particular nemesis. We live in a rural area so winters are traumatic for him due to the shooting that goes on regularly. When we first got him he didn't like car engines, radios, kettles, coffee grinders or machines: in short, anything that made a noise. I can only imagine how traumatic it had been for him stuck in kennels, shipped across the sea, and into another kennels.

When Billy first arrived he idolised Buck, our Doberman. Buck was a big, strong pup at the time, and none-too-impressed by having Billy hanging on his every dog word, slavishly following him about, sniffing wherever Buck sniffed, copying his movements: essentially, his very consistent and persistent shadow. Eventually, totally fed up and hemmed in, Buck did take matters into his own paws. Luckily, Billy lived to tell the tale, and although he still idolises Buck, he's a little more careful, now, about how he shows his adoration ...

We had never before cared for such a scared and nervous dog, and initially made many mistakes as we followed different advice about how to help Billy. We have tried herbal remedies, natural tablets, thunder/anxiety shirts, doggy calmers: nothing seems to really make an improvement for long. We are still working on it and learning lots.

Some may question the advisability of using Billy in sessions (he's not exactly the super-confident, laidback, take-everything-in-his-stride Labrador Retriever type!). However, the sessions have actually helped him as he does so love people and being cuddled and played with. That, despite his fears, he approaches each day with enthusiasm, gets out of his comfort zone, into his stretch zone, and occasionally into his panic zone (particularly when there is a loud noise) is something that people instantly relate to. Billy always wants to come out and play, and gets excited if he is going to 'work,' though does prefer outdoorsy activities, and finds classrooms a bit confining (who doesn't?) so we choose his work carefully

Billy offers the project and the people in it such a lot. When working alongside a man with severe anxiety, who hardly leaves the house, Billy can help him find the courage to make changes. Working out a plan to help Billy also helps people work out a plan to help themselves.

He is also a behaviour management tool with the teenagers, and it's not uncommon to hear one student say to another, "Be quiet, you're scaring Billy!" When we run anti-bullying workshops, Billy's great for building empathy, promoting discussion, demonstrating 'victim' body language, creating laughter, and establishing a strong connection as he cuddles up with people. He is our 'bravest' dog, and there's nothing that helps build empathy, and encourage kindness and gentleness, more than helping a dog!

Billy is also our 'attachment disorder' dog (a broad term intended to describe disorders of mood, behaviour, and social relationships arising from a failure to form normal attachments to primary care-giving figures in early life). Whatever is done for him, it is very often not enough. He has to be the centre of attention, and cannot bear it if another dog is stroked. He is never happier than when he is on someone's lap, pressed close against their heart, breathing in tandem. He also likes to roll on his back spontaneously, his long, Lurcher-derived legs frantically waving in the air.

He can also dance in a wobbly kind of way whilst standing on his back legs, which makes people giggle.

Helping Billy has been a major source of pride for many; especially one – Jodi – a large and usually intimidating girl, who had a reputation for fierceness, aggression, and violence. She could be like a whirlwind, and would regularly try and move others out of her way by bouncing against them so that they would sense her power. She was loud, harsh, and dominant in her approach to what she was and wasn't going to do; who she was going to do it with, and where and when. Any perceived insult resulted in a threat of physical violence. At the time she was the only girl in a special needs school full of boys – and she ruled the roost.

Somehow, Billy managed to pierce her armour, gaze adoringly at her, and become 'hers.' Being praised for how gentle, caring, and empathetic she was with him changed her approach to us, and sometimes others in the group. She did occasionally still threaten violence if she thought anyone might hurt Billy (not that anyone ever did), channeling all of her anger and fear into being protective. Jodi became a different person when she was with Billy, and we hope the memory of who she could be stayed with her forever.

Tyler, who we first met on K9 Action, and who now volunteers at our K9 Café (more about which later) was experiencing a lot of difficulties at school and home when we first met him. He loved all of the dogs, but had a special bond with Billy. When speaking at our conference (more about this later, too!) in front of lots of professionals – there's an increase in confidence for you – Tyler said: "I was in a dark place back then. Billy was my light; he showed me the way."

With careful management the project has helped Billy grow and develop enormously in confidence. In turn, Billy and his gorgeous golden eyes, have looked into the soul of many a young person, and found them to be kind and considerate. Most teenage boys would be reluctant to show kindness and care to another child, but working with Billy allows them the space to be soft and gentle, kind and considerate in a socially-acceptable way.

Taz

Taz, on the other hand, is a big, powerful dog, a Rottweiler-cross-Bull Mastiff who is not scared of anything or anyone. Handed in to the RSPCA by an elderly farmer who had five dogs he couldn't manage, Taz was pretty hard to handle to begin with. It took two people to walk him as he would drag them along (whilst at the RSPCA, he didn't get out of the kennel much as not many people could walk him). He had been at the RSPCA for nine months, and his situation was not looking hopeful.

When we first saw Taz, he was standing up on his hind legs, barking wildly at anyone and anything. He looked pretty intimidating; filled with frustration and boredom at his confinement. We didn't take Taz that day – we had only just taken in Billy, and did not want to introduce too many new dogs at once. However, I did find myself saying "If he's still with you in three months, let us know." Well, he was, and they did!

We worked very hard with Taz before we began using him in our programmes. We needed to be absolutely sure that he was safe, people-friendly, and extremely well trained. He has totally thrived with the work, and loves to learn.

Extremely confident, Taz presents a challenge of a different sort: all of the teenage boys want to work with Taz, but don't get to do so until they have 'graduated,' and proved that they can be trusted. What they don't usually realise is that Taz is actually easier to work with than most of the other dogs, because he's so laidback and happy to go along with most things, especially if a treat is involved!

Taz can also be what we call the Fun Police with the other dogs: stopping play that he hasn't instigated; taking toys away, and walking over the top of them if they are in his way. We exaggerate this a bit to use it as a demonstration of how unpopular bullies are. We might ask groups 'Who do you think is the bravest dog here?' and, initially, they always go for the biggest – Taz – and never think of Billy because he is so obviously nervous. The next question – 'So, who is the bravest: the dog who is not afraid of anything, or the dog who is afraid but keeps trying?' makes a hugely positive and powerful impact on children who have been bullied themselves: we have used this really successfully in groups of young carers, who are often bullied in school because of their circumstances.

Taz has also been a Blue Cross Education speaker dog and visited nurseries, Scout and Brownie groups, and schools. He is very patient and never tires of standing still while a queue of small children line up to stroke him. He is usually the same height or taller than many of them , so their faces are just at dribble height, so a dribble-cloth is an essential piece of equipment! Taz provides a fabulous example of the proper way to meet a dog, and the importance of having good manners around dogs.

One of Taz's other strengths is his ability to crowd-control and behaviour-manage. With Taz in the room, we've never had a student kick-off or show aggression: students have to prove themselves before he even turns up in class, so good behaviour is already established, and even those disposed to show a bit of bluff and bravado usually calm down when he arrives.

Hounds Who Heal

His major strength is also his gentleness and ability to be other than what he appears. He loves to get on laps and sleep and cuddle, but, at 9½ stone, that's a heavy lap dog! The students love it, though, as he offers so much non-judgemental love and acceptance. Students who are not accepted by the rest of the group for some reason, or who may be large in size, really relate well to Taz, and draw a huge amount of comfort from being around him.

Now, his age (nine-years-old) and size mean that the dreaded arthritis has set in a bit in his back legs, and he cannot carry out the more energetic activities. Nevertheless, he still comes out for a quiet, classroom-based session when he can, which makes him, and us, really happy.

Ruby

Ruby is a Staffie-cross, an ex-breeding dog, found on the streets, no microchip or collar and tag; unlaimed after seven days and on death row. We went along to see her, and she was a sorry sight: belly and teats hanging down to the ground after so many pups; back slung low, and very thin. She was sitting in the corner of her cage, her back pressed up against the bars, warming her tummy in the only patch of sunlight there was.

We were warned she wasn't so keen on men – Kev went down to stroke her and she growled at him. He stroked her tummy and she growled some more, as if to say 'Don't stop!'

We took her home. She cost £50.

The vet said she had been bred at every available opportunity in her relatively short life. We had her spayed (no more babies – I think she was more than a bit relieved!), after which she had a phantom pregnancy, and built a nest in the leather sofa: *right* inside it, having dug down to the wooden frame. We were a bit shocked, but it was getting past its best!

Ruby acts as our Staffie ambassador: I can't count the number of people with negative views of Staffies who have changed their mind after meeting her.

She is strong, loyal, affectionate, and also our dog body language demonstrator. She exudes joy! Or boredom, or whatever is happening for her at any given moment. She has strong opinions about many things, and will ensure you know what they are! She doesn't like the large, noisy, boisterous teenage groups, and will move to the back of the room if they become too loud. Eventually, one of the students will notice, and ask, "What's wrong with her?" Asked what they *think* is wrong, one will usually say, "Is it because we're acting silly?"

Ruby doesn't like shouting, and she doesn't like being teased. She also doesn't like the car, so only does work that is close by. She does like one-to-one training, lots of treats and clear instructions, lots of tummy rubbing and eye contact, and being out and about, and Ruby is the dog we use to talk about puppy farming, intensive breeding, breed-specific legislation, stereotyping, and how we should not judge books by their covers. The young people relate to this as they are often stereotyped for being young or looking a certain way. Discussing this in relation to the dogs makes it an easier subject to talk about; less threatening, somehow.

Ruby is a beautiful soul, who gazes into people's eyes, making them feel great, and trots along with them, busy and purposeful. As she has got older she has been less keen to do the work, preferring to be with me above anyone else. She does only the occasional session now as we totally respect her views and wishes.

Buck

Buck is our Doberman: regal, elegant, proud, keen to please ... well, over-keen to please sometimes ('Quick, quick, come here, come here, a leaf has blown off the tree!'). Buck is a very respectful dog, who would not lay a paw on anyone, and has had to be taught to rest his head on laps, because he previously thought that this was out of bounds.

He can also become very excitable and, well, yes, over-excitable at times, but looks great when he spins round and round in circles. He is a fabulous lead-walker, and the young people really enjoy the sense of control with such a large dog that this gives. Buck takes courage from their quiet approach – and also enjoys their sense of fun.

Kristian is a young man we worked with from a newly-opened school, and we were fortunate to be able to do all of the sessions in the RSPCA training room, taking along our own dogs, as well as working with some of the organisation's assessed dogs. The group was very diverse, and Kristian, at that time, was struggling with his new school, the academic work, changes that were happening to his body and mind with the onset of puberty, and also his ADHD. Erratic and easily upset, very few days went by when he didn't cry.

But he loved Buck, and Buck performed beautifully for him. Pointing at Buck, Kristian would say 'bang,' whereupon Buck would perform his only party-piece and obediently keel over, and this was such a source of constant delight to Kristian that we soon had to limit how many times a session he got to do it! Kristian's confidence grew, and we loved having him in the sessions. The day that Buck, in a harness, pulled Kristian along on his skateboard was a memorable one. Buck went awesomely fast until Kristian bailed out with a huge grin on his face!

Kristian's mum keeps in touch with us, and we know that he has grown into a lovely young man, studying at college, and with a girlfriend. She firmly believes that his sessions with us helped Kristian believe in himself, and made a huge difference to his life and development.

Dobermans also get a bad press, and to sense Buck's power, but also see how much he wants to please, is a very visual lesson for anyone working with him.

Izzy

Izzy is the latest addition to our group. We weren't really looking for a dog, but there she was. Another stray, found on the streets with no identification.

When her seven days and she was due to be euthanized, the good folks from a rescue came by and pulled her out. She is allegedly part Jack Russell, part Springer Spaniel, maybe Collie or Heeler? We don't really know, but think she's a blend of three or four working dogs, with traits from each. Small and completely hyperactive, Izzy is happy as long as she's working, working, working. Not all of the young people can handle her as she doesn't always do what they want easily enough, and some need the instant gratification of obedience, so do best with Cassie or Buck. Strangely, or maybe not so, it is often the girls who can persevere with Izzy.

Her difficulty in focusing for very long is one that many of our students can relate to. Her scattered mind, inability to tolerate boredom, and desire to be 'doing' usually wears out the most hyperactive student. As I describe how life might feel to Izzy, and what she struggles with, often a student will say "Do you think she has ADHD, Miss?" and we then discuss that, if the student has been diagnosed with ADHD or feels he may have it; what that means for him; what difficulties it may create for him at school, at home, or with making and keeping friends.

We discuss things that make Izzy happy, and what makes them happy. We talk about Izzy's good points and unique strengths and skills, and then theirs. We also discuss ways we can help her learn to relax and disengage from activities. Her need to have some quiet time, and her, and our need for balance between activity and rest. We also discuss diet and nutrition, and the effect this may have on our energy levels and brain function. When students become frustrated with her if she cannot do what they want, this opens a channel to discuss how their teachers/parents/carers may feel wth them.

Working with Izzy, Craig once said "You know, I get into trouble every day at school, in the afternoon. Apart from the days when I come to K9. I really want to get it right. It's not that I don't try. It's just impossible to get it right all day." This helped the school staff present see a different side to Craig, and view his behaviour in a different light, hopefully.

Izzy is also very affectionate if you can get her to keep still, and even experience trance-like relaxation if the moment is right, from which students can learn to relax themselves. She's also a fun dog who can provide lots of humour and playtime. She will run around for ages; her scentwork skills are awesome, and she can find things and people (even when in trees!). She's terrific at parkour-type activities, and loves to jump into things – including small shoeboxes. She makes the students laugh, and humour is often sadly missing from their lives.

And then there is Cassie. But she has a chapter all to herself.

12♥
Cassie: 'go-to-girl'

"She just is" – Alex, aged 12

Cassie is the canine star of the project. At nine years old, we have to accept that there will probably come a time when the project has to run without her, but I try not to think about that almost unbearable thought.

We first met Cassie, then called Kira, at an RSPCA rehoming centre when she was just twelve weeks old. She had been thrown from a moving car and found on the roadside, and was not yet out of the assessment section. She was a furry bundle of love and fun – a smaller-than-usual Shepherd-cross, completely unfazed by being in a kennel, and unaware that she was homeless – who approached everyone and everything with an open heart and a wagging tail. It was not at all hard to ask, "Can we have her, please?"

We had our home check and Cassie came to live with our current group of dogs, where she fitted in straight away. She was especially fond of playing with Beau, our beautiful golden Mastiff Shepherd, and Buck, the Doberman, who quickly became her best friend. Cassie and Buck were the same age, and had same black-and-tan colours, though he was twice her size!

Cassie played until she was exhausted, being rolled around and pounced on by two very large dogs. It made us nervous to watch sometimes, but they never did her any harm, and she developed a neat trick of rolling under the horse paddock fencing to get away from them when she'd grown tired. The fence was too low for them to follow her, and gave Cassie a chance to get her breath back, then pounce on them when they least expected it and reinstate the frantic, fast, furious back-and-forth playtime session. Then, sated at last, they would all flop, exhausted, to lie in the sun.

One thing we did not expect with Cassie was her dislike of any new dog. When, sadly, our dogs pass on, we usually get another sometime after. This is our positive spin on the fact that dogs live such short lives – even those who make it to eighteen still go far too soon for us – so we honour them by finding another shelter dog and offering him or her a home. But, whenever a new dog comes to live with us, Cassie hates them. She sits and glowers, usually from a perch on the sofa, and watches them, steely-eyed for weeks. 'I cannot believe they are staying! Are they really staying? Are they really necessary? Why are they here? When will they leave? Why don't they go away?' Then she stalks from the room, dissatisfaction oozing from every pore. Sometimes she comes around. Sometimes she doesn't.

But she does love people. Absolutely. She hasn't yet met one she doesn't love. She likes girls and boys (especially if they are wearing the treat bag), children, teenagers, and adults all equally. She forms close relationships with anyone, if they need her to. I have lost count of the number of young people who have wanted to take her home. When you ask them why they like her so much they can't explain. "She just *is*," said Alex.

Cassie was the one who cured seven-year-old Jessica of a dog phobia so bad that she used to run into the road in front of cars and fall down ditches to avoid dogs. After just three sessions with Cassie we were also able to introduce Taz the Mastiff, and Billy the Collie to Jessica!

Nineteen-year-old Emily was an autistic young woman who was frightened of dogs, and did not speak. It took two sessions with Cassie before she was holding the lead; three sessions before she was glowering at anyone who wanted to hold Cassie's lead, and four sessions before Emily was laughing as she ran along with Cassie trotting beside her. When Craig, who suffers from extreme anxiety, was brave to try and manage his fear of being with and talking to people by helping us run a fundraising stall, Cassie spent most of the day at his side, making the whole thing just about bearable for him.

When Cassie first worked on the project, she was young: the 'naughty' dog who jumped onto the pool table to chase balls and who stole food; who couldn't really concentrate for too long on her training. She was the perfect mirror to reflect back to our ADHD students what was happening for them, providing the opportunity for learning in a non-judgemental way. As Cassie grew older she matured into the work she did, and, although still fun and enthusiastic, she now presents a calmer picture, and is able to connect deeply with people in a very quiet way.

We have had to acquire another naughty dog (that's Izzy, in case you hadn't guessed!).

Cassie still becomes very excited when she sees young people, and she certainly knows what going to work is all about. She's interested in whatever they are doing, and watches while they play guitar or write in their workbooks. Youth club visits provide an opportunity to thoroughly investigate the floor for food, trotting up to those youngsters watching TV, and sitting on the sofa with them for a while, then going off to play football with whoever is outside. Cassie loves the training for school visits. She can perform some impressive tricks, such as the 'people jump' – jumping over a line of crouching students – and occasionally becomes vocal if she thinks she isn't getting enough attention.

When Cassie knows she's not required she simply lays quietly at people's feet, waiting for a hand to drop down and stroke her. When she works with people with autism or other communication difficulties, she is quietly persistent in asking for eye contact and interaction, and even if they drift off into their own world she doesn't give up. Meeting groups of young people out and about, Cassie trots over to them with a gleam in her eye: 'Are these mine to be with?' We often get strange looks, but usually there is at least one among the group who wants to meet her.

Cassie has a generosity of spirit; an ability to sense what people need, which she gives to them if she can. She knows when to nudge someone's hand, when to roll onto her back for a tummy rub, when to bark, as if to say 'Let's go and DO something,' when to quietly trot alongside someone trapped in their own internal world, and when to simply sit quietly by their side.

Cassie is the only one of our dogs who hasn't had formal training, or therapy dog training. She seems to intuitively know, care, and understand. She's not really best-suited to be a 'therapy' dog: she's far too active and self-opinionated to just sit about all the time! But her presence alone has been very therapeutic for many, and she has certainly taught us so many lessons, as well as provide insight into individuals and how they

Continued page 73

Top left: Buck connecting
(Courtesy Sylvie Gonera)

Above: Patch, the rescue dog,
connecting (Courtesy Sylvie Gonera)

Left: Billy connecting
(Courtesy Sylvie Gonera)

K9 Action
(Courtesy Ely Weekly News)

Top: Songwriting workshop with Maria Daines (Author collection)

Middle left: Taking a well-earned break to meet Maria's dog, Patch (Author collection)

Above: Meeting Brian and his assistance dog, Beasley, from Canine Partners (Author collection)

Left: Getting to know one another (Author collection)

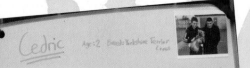

Julie and Ruby
(Author collection)

Cedric   Age: 2   Breed: Yorkshire Terrier Cross

Cedric has lots of energy and is easily excitable,
as well as loving attention, and liking to be around
people. He is a funny dog, and has a lot of
character. We know he is good with teenagers, as he
did 4 days of trainings on the Paws and Pals project,
where he learnt to sit, lay down, wait and to give
his paw. He also likes to be brushed, and is a
very friendly dog. We hope he finds
a good home!

Above: The young people made posters
to highlight how fabulous the dogs were
(Author collection)

Right: The ever-smiling Alfie (Courtesy Mark Fairhurst)

Participants and staff at Take the
Lead (Author collection)

Above, left There's nothing quite like puppy love. A deaf and blind Border Collie pup at the RSPCA (Author collection)

Above, right The Mayoress of Ely visits the K9 Café (Courtesy Mark Fairhurst)

Kieran making treat-foraging rugs at the Café (Author collection)

Summer holiday dog training fun at the K9 Café (Author collection)

Christian and Cassie training
(Author collection)

Billy: making that vital
connection, in his usual way
(Author collection)

Stacey and Ruby (Author collection)

Scooby looking apprehensive
on Day One (Author collection)

Top: Phil and Patsy (Courtesy Mark Fairhurst)

Above, left: Duke before we rescued him
(Author collection)

Above, right: Duke, in the care of
Greyhound Gap Rescue
(Courtesy Bilash Photography)

Left: Stampy at Glendee Rescue after a
pampering session (Courtesy Glendee Rescue)

Bandana dudes (Courtesy Kevin Willats)

... and with her rainbow
(Author collection)

Cassie collecting her Heroic Hound
Award, from Ben Fogle ...

Ruby
(courtesy Kevin Willats)

Chris and Billy
(courtesy Ryan Seville)

are feeling that we would never have realised without her. Cassie approaches everyone with the same easy, happy tail wag. She knows how to stay still, and be her companionable self without demanding too much. She knows how to offer encouragement, and can also command a situation if it's needed. She has an indefinable quality that provides students of all ages with motivation, a boost to their self-esteem, physical affection, love, and connection. She can get deep inside of people.

She just *is*.

13

Emotional Rescue

One Saturday I received a random phone call – "is this the lady who helps dogs?"

I could not have known, then, the full repercussions of answering 'yes' to this question (though I could hardly say no, anyway), with no idea of how many people this would involve, the broken hearts, and money spent ...

The dog in question was a small black terrier called Scooby, who had belonged to a drug addict and had lived a hard life. The woman on the phone was the drug addict's sister: her brother had died, and they didn't want his dog. Police and paramedics had broken into the house to try and resuscitate him; a job made all the more difficult because of Scooby's attempts to defend his comatose owner. When they later tried to remove the man's body, Scooby curled up on his chest in a tight little ball, and refused to move.

Of course I said we would help (already having found a place in my heart for this lost little dog), but knowing we were not in a position to keep him: I needed a real home for him and not just a little corner of my heart. The dog was 'difficult,' apparently, which ruled out all of the big centres which might have taken him. I phoned an independent rescue – one that I thought would give him a chance – which it agreed to do if we could deliver Scooby there. 'That was simple,' I thought, congratulating myself on how easy this rescue stuff was! We phoned the deceased man's neighbours, who said they would let us into the house, although, at that moment, Scooby had some visitors backed against the wall, and they couldn't leave the property. I smiled knowingly, wondering how such a little chap could be so intimidating.

We went there later that afternoon. Scooby was a cute,

button-nosed, scruffy little black terrier. A bevy of neighbours met us, including those we had spoken to on the phone earlier, a couple who used to help look after him when his owner went missing. The lady was sporting a rather large bandage as a result of Scooby sinking his teeth into her as she tried to stroke him. When we let Scooby out of the house and put a lead on him he was as sweet as can be: inspected our car, went for a walk, had a cuddle, accepted a treat. When we took him back into the house, however, his attitude changed; he appeared very worried, and sat on a jumper that belonged to his late owner, growling in a very agitated way.

There was no shortage of people keen to tell me about the dreadful existence that had been Scooby's: tied up in the garden for days on end while his owner went walkabout; left to fend for himself in the streets; not properly fed, and sometimes shut in the house for days. Lots of folk visited the house, and extensive drug and alcohol use was the norm, as well as occasional violence. No wonder the poor little mite was confused. Several people were hanging about, and offered to take him on, but, quite frankly, I wasn't convinced that they'd do a better job than Scooby's original owner.

Sunday morning arrived, and with it an email from the rescue to advise that it couldn't take in Scooby after all, as an emergency admittance meant there was no longer room ... Bemused and a bit concerned, we set off to collect Scooby, who happily came away with us. Now what?

It wasn't possible for Scooby to live with our pack of six as we had recently taken on some new dogs, whose behaviour was fairly unsettled and required close supervision. Adding in a dog I couldnt guarantee I could even touch at that point was simply not doable, and so, for a few days, he made his home in horses' feed room. He was dirty, smelly, and thin. His little button eyes were bright, but he was very confused and scared. He hadn't shown any aggressive signs, and, as my confidence about handling him grew, I became very attached. He needed neutering, and a local vet did us an amazing deal, throwing in a bath, hair and nail clip, dentistry check and clean-up for a reasonable fee (grateful thanks to Sally Anne at Cathedral View Companion Care Surgery, the first star in Scooby's tale).

Scooby was now looking much more presentable, and was ready to be passed on to a rescue, if I could find one, or go to an experienced home. My friend, Lindsay, who does lots of design and website work for K9, donated her month's wages to pay for his medical care (Lindsay, thank you – you're the next in Scooby's star list). Scooby was now living in the annexe next to our house, previously occupied by my dad, who had recently died after a long illness. The annexe had an adjoining door to our kitchen, so we popped in and out, and set up the computer there so he didn't feel too lonely. At night, he slept in a crate in our room. He loved his crate: it seemed to be the only place that he really felt secure.

Gradually, Scooby's quirks began to show. He had a thing about feet, and mens' feet in particular. He liked to sit on them, but if their owner tried to move, he would bite their feet – and when he bit he really meant it! Kev went into the annexe one day to use the computer without shoes or slippers. When he later tried to leave the room, Scooby sat on his feet, biting them every time he tried to move. Black dog, in a dark room: Kev couldn't even reach the light switch. I tried not to laugh when he indignantly told me about it later. I guess Kev was Scooby star number three, as he had to put up with a few attacks from the little fella. Scooby also became very attached to me, and didn't like anyone else coming close. It was difficult getting him to integrate with our group of dogs: he was fine but some of ours were not, and would not tolerate his attitude for very long! We simply had to find Scooby a home where someone could work with him, and provide the stability he needed.

The first home I found for him did not follow the instructions I provided, and Scooby bit someone's feet on day one: back he came. I realised I had no real knowledge of how to train him to make him more rehomeable – or the time to do it! Seriously worried now, I phoned my pal, Ross, a very skilled dog behaviourist, who put out the word about Scooby. His colleague, Vicky Lawes, who runs Lawes Paws Dog Training, offered to try Scooby at her home with her three dogs: a Rottweiler, and a couple of Shepherds. Vicky was at the point of completing some dog training courses, and thought it might be good to have a 'project' dog on which to test her new-found knowledge: thanks, Scooby stars, numbers four and five.

I drove to Ross' house in Oxfordshire, and met all of the resident dogs: Great Danes, a Shepherd, two Rottweilers, a Pomeranian, a Chi, and a JRT. Scooby trotted in like he owned the place, and, after a lovely lunch, off he went with Vicky. In my car I cried buckets; I could not believe how upset I was, even though I knew it was Scooby's best chance. Two miles down the road the car broke down. When I finally got home, five hours later, I was tired, dispirited, and weary, still missing the little dog like crazy.

It was three months before Vicky felt she could realistically try to re-home Scooby, who had bitten her son and her husband. (Scooby stars numbers six and seven). He also bit a stranger who rather foolishly tried to pick him up before Vicky had a chance to intervene. Vicky was desolate, as she had worked so hard, and thought she had eradicated this behaviour, as Scooby had reached a point where he behaved well, providing her firm and clear programme was followed. The trouble was, he looked so cute that people just wanted to cuddle him, but too much indiscriminate affection and fuss seemed to send him over the edge. Who could love him in the right way for *him*?

Vicky had got herself a lovely Spitz puppy called Fhiz: a delicate, fine-boned gorgeous boy (not originally intended to join the household whilst Scooby was there). When they found Fhiz with a broken leg, it seemed that Scooby might have been the cause ...

Ross, a regular contributor to *Dogs Today* magazine, wrote a long article about 'Da Scoob' as he called him, appealing for a home for him, and being honest about his difficulties. He received a grand total of three replies. Vicky felt Scooby had gone as far as he could with her, and needed a home to call his own – and soon. But it had to be a pretty special one ...

Vicky takes up the story.

"I had made all the usual approaches regarding re-homing a dog, and although everyone was very positive and helpful, nothing was forthcoming. It was early days, however, so I wasn't too worried. Shortly after my initial advert, a retired couple came into the shop, and we began chatting about dogs. They'd just lost one of their dogs, and had another elderly dog that was possibly not going to live for much longer. They'd always had rescue dogs, and not always the easiest characters, it seemed. I just knew when I met them that they were the couple for Scoob. I showed them his picture and told them the story, and, of course, how could they refuse?.

"I took Scooby to their home for an initial visit, and to talk through the specific training he had undergone to ensure that they were happy to take him on and continue with the training. All went well, so we set a date.

"I was relieved that I'd found a home for Scoob for many reasons, and was looking forward to leaving him with his new family to start his new life. I was also looking forward to getting my life back to normal. I delivered Scooby to his new home and, after a cuppa and a chat, left.

"I simply could not believe how miserable I felt leaving the little chap, and cried all the way home and into the next day. Gradually, over a few weeks, I became used to not having him around, but his going was a tear in my heart.

"A year or so has passed since then, and I see Scooby regularly. His owners love him to bits, and often bring him to see me. Scooby looks well and healthy, and is clearly happy. He's acquired a reputation for being a feisty little dog, and everyone knows him. His owners have just taken on another rescue, a little bitch, and the pair get on very well. Scooby is a lucky little dog who got his second chance, and is loving life."

So, 'Da Scoob' will live out his life with stars eight and nine, and his new doggy companion, number ten!

There's still a piece of my heart with Scooby's name on it, and, even as I write this, I feel very emotional.

Happiness and loss interwoven: dogs certainly teach you about that!

14♥
PAWS AND PALS

One of my favourite K9 programmes was Paws and Pals. Run with a very committed and skilled team of youth workers, this was an ambitious project involving a group of senior youth club members (aged 16 +) who would be our mentor group. They were already volunteer helpers at local youth clubs, and a more charming group of teenagers would be hard to find.

Because tthey were still a little shy and lacking in confidence, we provided training to be mentors, and some dog handling training, and then took referrals from social workers and youth workers for some younger children (aged 12 +), and spent five days at the RSPCA Centre working with the rescue dogs, training them every day. We'd undertake parallel group work sessions, set goals for the young people and dogs, identify obstacles, and create plans to overcome these, run teamwork activities, and spend lots of time with the dogs. The mentors paired with younger students and supported them through the process. This five-day section of the programme occurred over half-term one February: a big ask to give up most of their school holiday, but attendance was 100 per cent.

We were incredibly grateful to the staff at the RSPCA Block Fen Centre, who gave up their time each day to run dog training sessions, and provide us with a puppy party. They allowed us to occupy their training room for the week, set off the smoke alarm in the kitchen, and generally be a nuisance!

The students set goals for the dogs they were paired with, such as tricks or obedience exercises they wanted them to do. They mirrored this by setting themselves goals, too; the most common being speaking in public or speaking in front of people they did not know very well. By the end of the week the entire group had already achieved this objective! They also did some fantastic posters for the dogs to show how well they had done in their training, and how well they related to the young people.

The dogs we worked with included a YorkshireTerrier, several Staffies, and a Bull Terrier called Jackson, who was a real handful, but much loved and great fun. All the dogs we worked with that week were re-homed. The RSPCA staff were very pleased by how much difference the daily training sessions made to the dogs, and were clear the group's efforts had contributed to their being re-homed.

At completion of the project the group participated in a presentation evening where they all spoke on stage in front of parents, workers, and friends, and gave a PowerPoint presentation. They were presented with their certificates by the Mayor. I understand that all of these young people went on to make good career choices, and still talk about the project with great fondness.

VISIT HUBBLE AND HATTIE ON THE WEB: WWW.HUBBLEANDHATTIE.COM
HUBBLEANDHATTIE.BLOGSPOT.CO.UK
• DETAILS OF ALL BOOKS • SPECIAL OFFERS • NEWSLETTER • NEW BOOK NEWS

80

15♥

K9 CONFIDENCE AND K9 ACTION

'K9 Confidence' was the name of our first ever three-month-long programme for young people: twelve individuals who were experiencing significant school problems: some were being bullied, one had ADHD, one had Asperger's; one had autism. Some had serious difficulty focusing on anything for long, one could not stop talking, one never spoke, a couple were disaffected and disinterested in class, and one was aggressive with teachers: regularly throwing tantrums and storming out of class was the norm for both girls.

These were the kids who simply did not fit in at school – the non-achievers, the hostile: the apparently unreachable. One member described the group as the 'socially-awkward' set. Yet, they were all great, and, despite the diverse mix of young people, the group soon came together and began working well as a team.

The length of the programme – twelve weeks of three-hour sessions – meant we were able to develop good lesson plans and experiment with lots of new activities; and had the time to do classroom-based activities, hands-on work with the dogs, invite visiting speakers, and take trips. The programme took place during schooltime, so we always included plenty of structure and expectation of appropriate behaviour, as well as fun, active, fast-paced sessions. The group rose to the challenge and produced some excellent work. Most individuals showed substantial improvement in the school environment, too, which was very important.

The final week saw four of them on stage, presenting the project and their involvement in it to the entire school, which was a huge step for most of them. Feedback from teachers about the group's progress was hugely positive, which made a change to that mysterious thing 'attitude.' Gratifyingly, I received phone calls from parents to say how much difference the project had made to their sons and daughters.

When funding ran out, the school did not want us to stop working with their students, and so we continued the group in different formats on a voluntary basis, whilst applying for more funds. We did not want to abandon the young people, and especially at a point when they had done so well.

K9 Confidence eventually morphed into 'K9 Action,' an after-school club with a focus on making the world a better place for dogs and people. This group met in various ways for almost three years, and this was the longest period of time we were able to work with the same young people. For sure, we did not get paid for much of the work we did, though we did occasionally get lucky and received some funding, and the school found some money to keep us ticking over. Group membership changed here and there, but we retained enough of the original group to stay on as mentors to younger students.

This group achieved great things. They became really good dog handlers, winning rosettes at local dog shows, and getting to know our dogs really well. They learnt lots of new skills, and became involved in a wide range of fund raising projects, from manning market stalls and cake sales, to organising sponsored walks and completing a charity music CD. We ran a stall at *No Voice No Choice* – a fund raising day organised by DDA Watch, a not for profit organisation that campaigns for fair and effective dog laws. The young people learned about Breed Specific Legislation, and impressed many there by their responsible approach and attitude. On that day we were pitched next to Pen from Nowzad, and it was a real honour to actually meet Nowzad, the dog who started it all.

K9 Action temporarily morphed into the 'K9 Crew' for the writing, performing and recording of a music track called *One Dog, One Life*. The group wrote the lyrics after a day of watching various films and video clips, reading articles, and exploring the nature of man's relationship with dogs. They designed the sleeve notes, and sorted the graphics. We enlisted the help of our friend and animal activist Maria Daines, and her husband, Paul Killington. An accomplished songwriting and performing team, they use their music to raise awareness and funds for animal rights issues, and support a wide range of charities. We were lucky to have their skills and expertise onboard. When we began the project, around 13 young people wanted to be involved, but, by the time we came to the recording stage these had dwindled to just five, most of whom had said from the outset that they definitely did not want to be involved in the 'singing bit'! These young people still had very fragile self-esteem and low confidence, and it was up to Maria and Paul to convince them that they could do it!

It was a huge achievement when they finally completed the record and saw it professionally produced and packaged. The single was a great team effort; even if it didn't make the charts, the young people had pushed themselves way beyond their comfort zones to complete it.

During the course of the three years, we raised £1500 for our chosen charities Nowzad Dogs and Canine Partners; we'd been interviewed on radio; had featured in a local news programme, and had our photos in the local newspaper more times than I can remember, thanks to a very supportive, dog-loving reporter (thanks, Jordan!) We also won awards – in fact, we were finalists in the 2012 National Young Partners Awards, and got to attend a very swanky awards dinner in London, complete with posh frocks, candelabras, and chandeliers – the young people loved it! We were a winner in Cambridgeshire's 2013 Don't Waste Your Talent Youth Work Awards, and winner of the 2011 Innovation Category in East Cambridgeshire Business Awards. All-in-all, a very busy and rewarding time!

K9 Action eventually wound down when attendance grew poor as the young people, in their new-found confidence, moved on to other school activities (which was the point, after all!), or left school, and there weren't quite enough new members to make it work. It still remains one of the most successful programmes we have run for young people, of whom I have many fond memories.

I still see many of the students out and about, and they have turned into lovely young adults, some still with difficulties but most doing pretty good! They always seem happy to say hello, and, of course, always ask after the dogs.

16 ♥
TAKE THE LEAD

 I wondered if what we did with young people would work with adults, and couldn't think of a good reason why it wouldn't. The funding was becoming harder to find for working with young people, so I applied for European Social Fund Community Grant money to provide an employability skills training programme for those a long way from the job market. I got some help from Tim Cracknell, who headed our local support agency, the GET Group, to complete the bid, and, thanks to that, was amazingly, surprisingly, scarily successful! This allowed us to deliver two 12-week programmes in different areas: chance to source staff and venues, and really see how we could work with adults in a targeted way.

Finding the staff was interesting. We had money for sessional hours only, so it needed to be people who could work part-time for the duration of the programme. The job specification was pretty unusual: there aren't any other employability skills programmes involving dogs as far as I know! But some people still turned up for interview having not read the job spec or checked the website. Others gave fabulous interviews but did not accept the job; still others started but lasted just a few weeks when they discovered it was not what they thought. Regardless, I established an awesome, regular team in Kate and Dee, both of whom had impressive CVs, liked dogs and people, possessed enthusiasm, and bought new skills to the programme. Along with me and Kevin, this was our 'Take the Lead' Staff Team.

The first course was a huge learning curve, and we worked with a great bunch of people, several of whom have since become our core volunteers. Most had been out of work for between one and six years; we had a few young people who hadn't worked since leaving school, and several people nearing 60 who had experienced redundancies.

The second programme was totally different again. The participants were equally awesome but experiencing different barriers to moving forward, and we were humbled by the difficulties they faced on a daily basis. We had referrals for people who hadn't had a job in twenty years, people with learning difficulties, those enduring mental health problems, and recovering from addictions. Some were too ill to complete the course. Those who did made amazing progress: an inspirational group who bonded really well and supported each other.

By the third course, we were well into the swing of things; we had a clearer idea of what we could realistically achieve, and a structured but flexible programme in place. We had also learnt that one session a week over 12 weeks was in no way long enough to help those being referred to us. 'Move on' placements (other, less specialised community placements) weren't easy to come by, and it always felt as if we were abandoning our people at the end of the course. Keeping in touch was a bit ad-hoc, and we realised we had let some participants 'crash.' Part of this might have been due to our lack of knowledge of the system and other agencies that could help, but, for many, it was the lack of appropriate volunteering or training opportunities, especially if someone wanted to work with animals. Volunteering in a charity shop – the usual offer – may be very helpful for some, but does not work for everyone.

So, what to do?

One solution was to set up our own projects to provide our own opportunities: K9 Walk Support and The K9 Cafe were born! We are now looking for further funding for these projects, and have a regular group of Take the Lead graduates who volunteer in a variety of roles. Sometimes, it can be nine months before members are ready to come back to us to make the changes talked about in the sessions, so we need to be around in the long term in order to help people make those changes.

After running the programme a total of three times we know that it works. We've written a training programme that we believe is unique and effective (we've had some admirable results), and, at the time of writing, we are currently bidding for more European Social Funding help. We're also looking for corporate partners to possibly help with funding the programme in different localities.

Of course, the dogs are at the centre of this programme. Supporting, providing feedback, guiding, giving those who struggle to communicate something to communicate about, be this talking about their own dogs, our project dogs, or dogs we meet at rescue centres, those we see on TV, in the newspapers, or on Facebook. There is a common love and interest, a safe and neutral talking point, a way of getting to know each other, testing the ground without having to share personal information. The dogs provide a distraction, a paw to hold, a head to pat. I once watched Billy work his way round a table full of people, receiving a pat from each of them. Cassie once sat with a chap who was really anxious for almost the whole of a bric-a-brac sale that we did, popping back to me now and again as if to say 'Is this okay?' Taz helps people improve their body language, posture and assertiveness by being lead-walked in a public park.

We maintain as much contact with as many Take the Lead graduates as we can, and are inspired by their continued progress and motivation.

17♥
LATEST MOVES

Out of Take the Lead emerged K9 Walk Support and The K9 Café, because it seemed so wrong to drop people at the end of a project, when they had often only just begun their journey. These two projects have become the cornerstones of our adult services, and are addressing a huge need for many. Unfortunately, at the moment, we have very little funding for either, but we are carrying on, regardless. Both projects provide further opportunities to socialise, give and receive support, and build on volunteering and other skills, as well as put something useful back into the local community.

K9 Walk Support is a volunteer project providing dog walking services for those whose dogs need exercise, but who are unable to give them everything they need due to illness or disability, and are unable to afford regular dog walkers. It is definitely a win-win deal: our volunteers have a reason to get out of the house, learn new skills, be with dogs, and help others whilst helping themselves; dogs receive the regular exercise they need, and their owners feel connected to K9, and receive the help *they* need. Phil, who features elsewhere in the book, is the main co-ordinator for this project, supported by me. We have about six volunteers, and so far are providing 15 hours' dog walking each week. We hope to expand this and access funding for dog-related volunteer training, as well as more generic training, such as team leading.

TALKING DOGS; TALKING LIVES: THE K9 CAFE
The K9 Café in Ely is more a café for people than dogs, but we get a lot of doggy visitors, too. Initially envisaged as a drop-in for people we already knew, or who might be interested in getting

involved with us in some way, it has developed into a meeting place for all kinds of lovely folk. Our project dogs are always there to interact with, and the café is run by our volunteers; most notably John.

John never used to leave his house much, and experiences bouts of depression, but as a Take the Lead 'graduate,' he is now responsible for running the café. He does all the setting up and drink-making, welcoming new visitors, and clearing up afterwards.

We are fortunate to have a venue that supports us with a low rent that we can afford, is happy to allow dogs in, and loves to see it go well. John now has a set of equally able volunteers who help. Visitors include Malcolm on his motability scooter and his Labradoodle, Charlie; Sandy with her lovely dogs, including the very elegant Alfie with three legs. We have played host to Cockapoos, Jack Russells, Staffies, Shepherds, a Puglet, Golden Retrievers, Lurchers, and a Staffy-cross-Jack Russell who comes in to see another volunteer, Steve, for some dog training assistance. A local lady with substantial learning difficulties comes along, and her challenging behaviour gives volunteers a chance to develop strategies for dealing with difficult people! She can be very persistent about having things for free! We have a number of older people with dementia with their dogs, and the café provides regular, safe and consistently friendly support. We keep an eye out for each other and the atmosphere is very welcoming.

We are all learning to be kind and empathetic, but clear and firm, and convey what we mean very clearly and simply. The volunteers are all learning new skills. We have carers bringing in people in wheelchairs for a doggy cuddle and a chat; the Mayoress of Ely has dropped by, and the café is fast developing as a hub for some of our other work. We have held a successful CD, DVD and jewellery sale to raise money for venue hire. We have a dog groomer who comes along regularly to provide free advice and nail trims (for dogs, that is, not people!), and the tables now sport a funky selection of doggy-themed tablecloths, thanks to a volunteer who can sew. Our friend and supporter Mark Fairhurst visits and takes fabulous photos, many of which feature in this book, and promotes us in any way he can, for which we are very grateful. Ryan from Neighbourly.com (the first social network to connect local projects and community needs with companies ready to help with funds and volunteering programmes) shot an awesome video, which you can see on our website.

One of the positive aspects of not having funding is that we do not have to reach a target, and the projects can develop and evolve in a natural way. We can provide a place for people to just 'be,' and not have to tick 'doing' or 'outcome' boxes. Coincidentally, lots of 'doing' actually gets done!

Looking around some days I am inspired by the variety of people who attend, and how we are learning to interact with each other. I am not sure there are many projects that deal with mental health, learning difficulties, social anxiety, young, old, and everything in-between in such an inclusive and friendly way. And, of course, the dogs!

We would, of course, love to have our own venue; to be able to open more often, and widen the range of activities we can offer and become involved in. A Dog Hub pop-up would be awesome. Perhaps we could offer free help and advice, maybe some dog-sitting and dog-walking services, and also advertise and promote local rescues and dogs who need homes, along with promoting responsible dog ownership. All things to think about – who knows?

We will continue to fund raise, let the K9 Café expand naturally, and look forward to meeting lots of new people and dogs. I am now helping other community groups set up their own version of the K9 Café – the first step towards K9 Cafés all across the country! What a dream!

Visit Hubble and Hattie on the web: www.hubbleandhattie.com
hubbleandhattie.blogspot.co.uk
· Details of all books · Special offers · Newsletter · New book news

18♥
Charlie

 The first time I saw Charlie, he was a small, thin, vulnerable 13-year-old, wrapped in a parka coat, twitchy and shy, unable to communicate. The last time I saw him he was 16, almost six feet tall, able to express himself, and inciting other youngsters into anti-social and dangerous behaviour, in trouble with the police and losing his grip.

Charlie had formed a close bond with Cassie, and I had hoped that, between us, we could halt the seemingly inevitable decline that loomed. It was a false hope. Charlie had spent most of his life in and out of care, victim of a noxious cocktail of violence and neglect at home. He was expelled from schools, his mum was dying of cancer, he had a dad he didn't know, a step-dad he didn't like, and a scattering of brothers dotted about two counties he never saw. He'd been let out of care when we first met him, so that he could go home and be with his mother as she deteriorated, and then died. During her illness, she was wheelchair-bound, covered in heavy bandages for the weeping sores across her body. Even before she became ill, she was, apparently, never an easy or approachable woman.

Charlie attended a special unit for children with emotional and behavioural problems. He was a kid in the 'syndrome mix' – ADHD, OCD, mood disorders, and potentially on the autistic spectrum ... his behaviour was a whirlwind of volatile mood swings, counting, fixations and obsessions. He counted cracks in the pavement while making sure he didn't step on them; he counted fence posts on picket fences and windows in the youth club. His obsessions included keys of any description, pickled onions, and gherkins. As his fixations changed regularly and with little warning, the large jar of gherkins you had bought for him was soon redundant as that fixation turned to something new – chocolate spread or cheese, say.

Charlie had little grasp of reality, so making sense of the stories he told was always challenging. He tried very hard to make sense of what was happening to him, but didn't always have the words to describe or explain in a way we could understand. He would describe himself as 'scattered' when he could not concentrate or sit still.

Once he started working with us, he developed another fixation: Cassie, our Shepherd-cross whom he loved. His own dogs had died after fighting each other, and, coming soon after his mum's death, this was hard for him to deal with. Cassie became incredibly special to Charlie, and he had photos of her plastered all over his bedroom. The other students at the school knew Cassie was 'his' dog, and although they were 'allowed' to work with her, they knew he was always watching them! He felt that no one ever took notice of him or gave him what he wanted and needed, and he once said of Cassie, "I can't believe that anything that perfect, that wonderful, cares about me, and does what I ask her to do. It's unbelievable!! I've never experienced anything like it."

To make it even better, Cassie really loved Charlie, too. She sensed when he felt sad, and went to him, gave him a kiss, and asked for a cuddle. She knew when he wanted to run and play. She followed him everywhere, and made him feel loved, cared for, and cared about. He taught her new tricks and games, and they had absolute fun. When he fell into an exhausted dozing state, Cassie would lay with him. Once time when she went over to him and licked him, he said, "Thank you, Cassie, for noticing I am sad," which opened the door for me to ask if he wanted to talk about his sadness.

Cassie and Charlie hadn't seen each other for three weeks during the long Easter school holidays, and Charlie was worried: would she remember him? Would she still like him? He'd spent most of his holiday in his room, gazing at her pictures, so that she greeted him every morning, and he said goodnight to her every night before he went to sleep. Hers was the first face he saw each morning.

We drove to the country park where we met and walked with Charlie. Cassie spotted him across the car park, and ran to him, tail wagging, whining in anticipation and excitement; they rolled around on the floor together with Charlie crying with joy.

Cassie was jumping on him and yipping and licking him. I had to turn away and brush away the tears.

As Charlie grew older and diagnoses kept coming, he was put on medication, and then adolescence hit with a vengeance. He became taller and taller each day, it seemed, and was suddenly no longer a little boy. After his mum's death, Charlie was living with a step-dad he hated, who was probably struggling with his own issues at the time. Our efforts to provide extra support during the long school holidays were not successful, and it seemed as if his step-dad was sabotaging any way of helping Charlie. School staff were becoming frustrated, and Charlie kept asking if he could come and live with us, or with previous foster carers. He was struggling to cope with how his medication made him feel, and often refused to take it. When he returned to school at the end of the summer holidays, he was behaving very differently ...

His conduct further deteriorated, and be began to abscond from school, taking younger kids with him. He still loved Cassie and wanted to take her somewhere alone, but his behaviour was so erratic we could not allow that. His relationship with Cassie and us no longer provided any leverage for good behaviour in the sessions. He grew more distant; the way he acted around the dogs was not safe, and he tried to abscond with Cassie and another boy. Other lessons were also going badly wrong for him. The decision was made to suspend him from our sessions for a while in the hope that this would provide a wake-up call. It couldn't and it didn't. Charlie grew even more distant until, eventually, he said he no longer wanted to come to sessions. We simply could not engage him in anything, and, frustratingly, sadly, unimaginably, we lost him. I was devastated.

How fantastic it would be to say that, eventually, and in part due to his relationship with Cassie, Charlie was able to shape up, turn his life around, and go on to be a successful and happy person. But the truth is I have absolutely no idea where he is or what happened to him, although my intuition tells me it is unlikely to be good.

What I do know, though, is that, for almost three years, Cassie provided Charlie with a reason to get up, and a reason to care; non-judgemental acceptance and love, physical affection, enthusiasm, and hours of fun. She loved him in a way that no one else seemed to, and at a time in his life when he so badly needed it. For my part, I wanted more for him than that, and hoped that, as he learnt what it felt like to give and receive that love with Cassie, he would open up, and there would (should!) be a person caring for him in the same way. I can only hope that some of what he learned and felt through being with Cassie will filter through into other experiences, other relationships, somewhere in his future.

'D'

D was small for his 12 years, with a bright mop of ginger hair, lots of freckles, ADHD, and Tourettes. He had left most of the male members of his family in the south of England, and come to Cambridgeshire to live with his mum and sister. His mum had cancer, and wanted to be closer to better hospital facilities.

D entered his special school like a whirlwind. He hit just about everyone who crossed him, and was soon barred from most lessons because teachers refused to have him in their classes. He was fiery, over-reactive, and volatile. He was also angry, scared, lost, lonely, and afraid.

He said he'd come and do the 'dog stuff.' He liked dogs and was good with them; unfortunately, he was not always so good with other students. The first time he started a fight, he smacked another student around the back of the head without warning, and they both fell to the floor, rolling around. Cassie and Billy began to circle them, wondering what was happening, and whether they should join in the fun, bark, or jump in to stop it. I got the dogs out of the room pronto. The skirmish broke up, and D was banned for a few weeks.

Not much was working for him at school, however, and we allowed him to return. We had already grown fond of him, and could see that he had a lot to offer. He loved the dogs, and around them was already becoming responsible and the best he could be.

A few weeks later, D felt he had justification for hitting another student, but, knowing that he shouldn't, he threw a chair about instead and ran out of the room, making threats.

This time, we decided he would not come back, as

there was a chance the situation could kick off again, and be dangerous for all concerned. We telephoned the school, and asked that someone come and collect D. Before this could happen, D came to the door and asked to be let in. I refused. He said he needed to go to the toilet. I shrugged. After a few mnore comments back-and-forth, he said, "You can't do this." "Watch me," said I, taking a gamble that he wouldn't attack me, or try and force his way in. D seemed genuinely bemused that I was sticking to my guns, and eventually left with the school staff when they arrived to collect him.

The next time we saw D in school he apologised, and asked if he could come back. We pretended to deliberate, and made sure he knew how he had to behave. He returned to us and became a very special student: good with the dogs and helpful to other students (unless they really ticked him off!) But now he had learned to walk away from conflict; he became a 'helper.' He asked questions, was up for activities and written work, and gained a tremendous amount from physical contact with the dogs – he loved the cuddles.

D's home life was awful, although he stayed in all the time. He had no friends locally, as he was sent to a school a long way from home, so did not leave the house once home from school. His mum was sick, and his sister viewed him like most big sisters view their little brothers! Our session was scheduled last on a Friday afternoon, and we felt good about sending him home for the weekend a little less alone, a little more calm and relaxed, a little more sympathetic to his mum's condition, and maybe with a dash of self-belief.

One session we had was particularly memorable, when he was doing agility with three different dogs. Here was a boy who appeared unable to control his temper or the way he communicated, who had little patience, and who seemed to antagonise everyone he came into contact with, working with three dogs to get them round the agility course. The dogs were all very different, and we watched in awe as D used a firm, authoritative, but motivating and happy voice along with assertive body language with Taz; a kind, encouraging, playful voice and lowered, precise, slow, body language with Billy, and a high energy 'let's go do it' voice and body posture with Cassie. He was able to adapt and monitor what he was doing to get the best from each dog. He coped with the fact that not all three of them did a perfect round, and praised Billy for what he had managed to do. He walked back to us with a look on his face that said 'Should I be pleased with that, or not?' We heaped praise upon his head, and were careful to explain to him exactly what he'd done that was so impressive. He beamed from ear to ear whilst blushing furiously. I can remember telling him that there were very few students we worked with who could do what he had just done, and lots of adults who couldn't do it, either!

The dogs usually attachéd themselves to particular students, and Izzy opted for D. One day, as he was working with her (she is a little bit ADHD herself, something not lost on D), I wondered aloud if we could help her relax. D was not hopeful. I asked if he'd be prepared to give it a go, and we settled on the floor, with Izzy wriggling. I asked him to breathe deeply and slowly, and really focus on moving his hands slowly, timing his movements with his breathing. I slowed my breathing, too, and began to repeat a mantra in a lowered voice. Izzy resisted intially, but then relaxed,, breathing deeply herself and pushing herself into D's hands. His eyes were almost closed, and the atmosphere in the room was hypnotic. 'How do you feel?' I asked afterwards, and although he really had no words to describe how he felt, he was obviously profoundly moved by the experience. He did say he felt 'strange,' as he had not felt that calm before, but was really pleased that Izzy had trusted him so completely that he was able to calm and relax her.

It was a moving experience for me, too, knowing that this young boy had never had a moment's calm in his everyday life, his head continously a-jangle with noise, thoughts, worries, anxieties and anger. He was always very clear that the only reason he went to school was to come to K9, and I had no reason to disbelieve him.

We left D at the end of the school summer term knowing that the next six weeks his life would be 'rubbish' (his word), but hoping we had given him enough self-belief to maybe make new friends. I tell the story of the school that D attended elsewhere, so, for now, let's just say that we were not asked to go back after the holidays, and never had a chance to say goodbye.

I learnt from another student that D returned to school the first day of the autumn term, but, learning that we were not going to be part of his curriculum, did not return again.

We have no idea what happened to him.

20

PHILLIP

The first time we met Phillip he ran along in front of us, repeatedly hitting himself on the head and muttering "I'm f***ing useless, I am, I'm f***ing useless, I am." He was 12, with significant learning difficulties, that meant he struggled to make sense of many things, and was pretty much a walking disaster area. If Philip cooked, it resulted in more ingredients over him, the table, the cooker, and the floor, than in the dish. If he actually managed to cook anything, he usually dropped it before he had chance to eat it – and usually in the only puddle around.

Eating in public was always interesting: he spilt food and drink on tables, floors, and clothes. His shirts were constantly coming untucked from his trousers, and he showed off his belly at any and every opportunity.

Phillip's family constantly told him how useless he was, and at home he was banned from doing most things due to the resulting disaster zone scenario. He had also been told that he did not need to learn to read and write, as it wasn't at all necessary to do so.

Phillip truly believed he was useless, and it took him a long while to trust us, talk openly, and learn to join in and have fun. When he was happy, he was the greatest fun to be with; we were always smiling. Driving away from school, he would lean out of the window, singing along to the radio or a CD, shouting "'ello mate" to pedestrians or other car drivers, and muttering "w*****s" under his breath if they didn't reciprocate. He greeted all strangers like long-lost friends, and was deeply wounded if they did not respond with equal enthusiasm; often they did not even realise he had spoken to them, and wondered why a boy was looking at them and muttering profanities under his breath!

One time, we had given Phillip a worksheet, on which he had to list four things you should not do around an unknown dog. We place strong emphasis on safety, so usually get replies along the lines of 'do not run,' or 'do not approach a dog you don't know.'

Phillip's answers were –

TOUCH

SHOUT

SPIT

Time with Philip was always fun, albeit a little scary in the event that he really upset someone. Underneath all the fun, though, was a boy with limited opportunities, a lack of family support ... and an uncertain future.

21
AMBER

 mber came to one of the group projects that we ran with a local youth work team. She was around 17, and was one of the young mentors who was going to work with the dogs and mentor a younger student, too. Amber was extraordinarily beautiful, with gorgeous skin, black hair, and large, dark eyes. You could tell that bad things had happened to Amber: she looked as if she carried a heavy sadness in her heart. She had not attended school or college in years, and also did not speak much. Well, she didn't really speak at all, actually.

We wondered how Amber would be able to mentor someone if she could not even speak to them. When she worked with our dogs she was wooden and stiff, avoiding eye contact, and not talking to them, so that they struggled to know what she wanted them to do. But she kept on coming, and we kept on encouraging her, and hoping.

One week of the project comprised five days at our local RSPCA Centre, where we were so lucky to be working with the rescue dogs. The programme focused on helping the dogs become adopted, and helping the students gain confidence and new skills. Goal setting, initially for the dogs, was a crucial part of the programme, and group members then set their own goals. Most focused on speaking in public, and in front of people they did not know. We set lots of little presentations for the group, and they had to work as a team. We wondered whether Amber would attend the whole programme, as she obviously found it difficult and painful, although that she kept coming back was testament to her inner strength.

Initially, Amber was her usual rigid and stiff self, speaking little in the presentations, her fists clenched in fear; her body tight. The younger student she was due to mentor had not turned up, which was somewhat of a relief. Amber was paired with Ruby, a little black Staffie pup who had been overlooked at the shelter for a long while.

At some point during those five days – I'm not sure quite when it was, but it was pretty quick – Amber opened her heart and Ruby jumped right in, quickly followed by a couple of squirming blind and deaf Border Collie pups who amazed and inspired with their ability to chase a rolling ball, with neither sight nor sound to guide them. By this time, Amber was smiling an open, luminous beam and moving more easily, noticeably looser and relaxed. Her eyes were open and she was making eye contact. When she gave presentations, her hands were no longer clenched, and she was talking to others.

When we finished the week, and Amber had to say goodbye to Ruby, there were lots tears and deep sadness. I worried her heart would close again.

A month after the programme ended, we held a big presentation evening. The entire group, including Amber, conquered their fears to get up on a stage in front of families, friends, teachers, social workers, and those involved in the programme to make speeches and accept awards from the Mayor, and they presented the workers with gifts.

Three months later, Amber enrolled at college on an animal care course. As far as I know she is still studying, and the workers who see her tell me she is doing great.

I think she has travelled on with an open heart ...

22♥
STACEY

Stacey telephoned to ask if she could come on a Take the Lead course, her voice a whisper. One of her workers had suggested she do it as she loved dogs. I explained a bit about the course but wasn't sure it was really sinking in, although she said she would come anyhow. I offered her a lift. She asked if her dad could come with her. Sure, I said – whatever it takes.

The first week she hardly spoke, and, when she did, it was in a whisper. I wondered if she would come again. After the first week she didn't need her dad to come along, although I still collected her. After about four weeks she was coming by herself, and joining in the discussions. By week eight, she was actively supporting other people in the group.

On the final week – week 12 – she zapped into the room on her roller blades, long hair flying, with a huge smile on her face. The changes in Stacey were huge. She is a beautiful young woman who comes alive when she is working with dogs: she finds a new level of something inside her that she lets out. She still has a way to go, and her opportunities are limited, given her circumstances and what she wants to do. But when the right opportunity comes along, I am confident she will grab it with both hands.

This is her letter to us at the end of the programme.

Dear K9
I wanted to write to you to let you know how you have helped me with the Take the Lead Programme.
The first time I came to the group I felt scared of meeting new people. I didn't really know if I would get anything out of it. It was really hard for me to come to the group, and I struggled

so much just to get out of the house that I had to bring my Dad with me. I was so worried that something would go wrong and I would panic. Dad being with me that first time made it a bit easier.

Chris used to have to pick me up just to get me to the group. After the first session, though, dad didn't need to come with me anymore. I knew the dogs would be there, and I planned with the group leaders that I could take a break if I panicked. The staff let me walk the dogs into the group so I didn't have to walk in on my own.

The dogs helped me with my confidence. We did some work with body language, and how using it right helps the dogs. The dogs didn't judge me if I made a mistake; it didn't matter, I could just be myself.

Take the Lead has encouraged me to go for what I want in life. I didn't think I was going to go anywhere. I didn't have the motivation and I didn't even see the point in living before the group.

Take the Lead has given me the confidence to talk to people again. Before I started the course I couldn't talk to anyone and I would avoid people. The group helped me meet new people and talk to them.

Knowing that the other people in the group were feeling similar to me helped me feel more comfortable talking to them, and sharing my feelings with them. It has helped me build trust in people, not just in the group, but in my life generally.

I have been looking at working with dogs in the future, and doing some voluntary work to help me build confidence more and more.

Take the Lead has given me a purpose. Since coming to the group I have found the skills and confidence to move forward with my life. I used to isolate myself because I was terrified of meeting new people. Now I look forward to the next step on my journey.

Thank you
Stacey

23.♥
PHIL

The first thing I noticed about Phil when we met was his hair. It was very long, and a beautiful golden-ginger colour that glowed. I had no choice but to notice it, really, as most of his face was covered by it, and he peered out in my general direction from behind its shiny curtain. The second thing I noticed was that he was tall – very tall, in fact, at 6 feet, 8½ inches. Ginger hair and very tall. Bound to have had some difficulties with that at school, I thought.

Phil had been referred to me by a local Connexions (a UK governmental information, advice, guidance and support service for young people aged thirteen to nineteen) worker. He was 17 and doing nothing. I didn't really know why I had agreed to visit him as we had no funding, Connexions was not in a position to pay for anything, and I had vowed not to do any more voluntary stuff: the project was supposed to provide my living, and I wasn't well off enough to work for nothing.

Nevertheless, here I was, and Phil seemed like a nice young man with loads of potential, although obviously depressed, and so hardly ever left the house. Behind the hair you could see the young man he *could* become, given the right opportunities, and a chance to conquer his depression. And a dog. He wanted a dog. I could see that he needed a dog, and thought there was probably a dog out there somewhere who needed Phil just as badly. Despite there being no funding to work with Phil, I couldn't turn my back on him – I simply had to be creative about how to get him involved in something. I didn't think it would help him to be with a group of young people, so it had to be something on an individual basis.

Kev and I were volunteers at our local RSPCA, and would

walk its dogs whenever we could. We decided to take Phil along a couple of times. He loved it, and slowly began to talk a little more. I decided to have Phil help me and Tyler (that's Tyler of the "I am one of the socially-awkward group" fame) run some fund raising tombola stalls. I thought they might get on okay as they both liked all the geeky gaming computer stuff that I don't understand, and would have something to talk about. I also knew this would be hard for Phil, and was a big ask: to be in public, talking to strangers, handling money, trying to persuade people to buy tickets. I find it difficult, too!

Phil came along to the event and, initially, was very quiet, though began to communicate the more we went out. I have a photo of Phil, Tyler, Cassie, and an elderly couple, Mr and Mrs S (who had phoned me out of nowhere and asked me to help them get a dog from a shelter, which I did), sitting on the grass eating ice cream that Mrs S had bought. She had dementia and her husband was often poorly but they had adopted an gorgeous little elderly Yorkie who was the light of their lives. I remember looking at all of them, and thinking 'what an unlikely group,' and how wonderful that complete strangers can connect because of dogs. I still feel very emotional thinking about that time, which, for me, symbolised all I had hoped to achieve with the K9 Project.

Phil managed to persuade his parents to let him get his own dog from the RSPCA – a lovely Malinois Shepherd mix called Patsy, extensively used for breeding in a very bad fighting/breeding dog establishment. The RSPCA had seized lots of dogs there, and Patsy was in a bad way. Phil has worked wonders with her, however, and she is now a beautifully calm, friendly dog who we can use on some K9 Project programmes, and a credit to herself and to Phil.

I can also tell you that Phil is almost unrecognisable, now, compared to the young man I met way back then. His hair is short – a sure sign that he is feeling more confident about facing the world without the need to hide behind something. But I will let him tell you in his own words what kind of a journey his has been.

"In 2010 I started sixth form, but, unbeknown to me, the college I chose (I'd gone to the secondary school that shares its grounds, and I lived around the corner from it) was one of the worst in the area. A series of events leading to severe bullying, including death threats, unfolded at this school, and were not helped by the less-than-caring staff, some of whom even joined in with the bullying! That's not to say that none of the staff cared – a couple did – but they were limited in how much they could help because of their position in the school.

"These events resulted in deep depression, and I felt as though I couldn't carry on. I thought that the few people who still talked to me only did so out of pity. I hated going to college, to the point where most days I simply wouldn't. My parents have always been strict about my attendance, but even they agreed on many days that I was better off at home. I realised something was wrong when it occurred to me how regular the thought of death, or specifically suicide, would float around in my head, and how much of a release I was beginning to feel it would be.

"I went to the doctor and he sent me to a youth service called Connexions for free counselling. It helped a bit; gave me a sounding board. I could talk to someone completely neutral and just let everything go. I had two counsellors I'll never forget: first Suzanne and then Kate after Suzanne was transferred. It was Kate who put me in touch with Chris Kent, founder of the K9 Project, in late 2011, and around the time I'd been getting at my parents to let me have a dog.

"Chris and the K9 Project helped me a lot. I started there by helping with fund raising days, which forced me to leave the house (a rarity at this point). It was terrifying at first, but Chris made sure I knew that I only had to say the word and I could go home, and I didn't have to do anything I didn't want to.

"After I'd been out with them a couple of times I began to look forward to it., I was getting to know Chris and her husband, Kev, very well, and the work we did boosted my confidence, if only a little, despite my never talking to anyone. I stayed close to Chris' dogs, who kept me calm. To begin with, if anyone approached me to talk, I'd just point to Chris and hold onto the dogs. It wasn't much, but it was more than I'd done in a long time.

"In early 2012 I got my own dog, Patsy, and that was mainly down to my parents seeing how much being around the K9 Project dogs had helped me. Patsy helped me a lot with going outside: because of her, I had to leave the house at least twice a day, whether or not I liked it. I still struggled to talk to people, and when you're a dog walker that can make life difficult as a lot of people will approach you.

"Later on in 2012 I took part in the K9 Project's employability skills programme, Take the Lead. I wasn't very confident about meeting so many new people, and honestly didn't think I'd go more than once. What I forgot was that everyone else was in the same boat, and we had a real connection – we all loved dogs. It was a talking point, and the dogs were there with us to keep us calm. It also helped that my only two friends from college had joined, too, so I was able to walk in, knowing I already knew at least three people, including Chris.

"Over the 12 weeks we got to know each other very well: learning each others' strengths and weaknesses, teaching each other to build on what we knew, and boosting our confidence. We were a small group from all walks of life: 18-year-olds fresh out of college; single parents; married couples, and even a couple of people in their fifties with decades of work experience. We ended with a charity day, which we set up ourselves, and, while it wasn't exactly a success, the point was that, between us we'd done it; proven to ourselves that we were capable.

"While my confidence had improved a lot, thanks to K9, we still weren't done. In the summer of 2013 I helped out with running sessions in another Take the Lead course, which was just what I needed to push my confidence to right where it needed to be. By the end, I could talk to strangers confidently, voice my opinion, and leave the house when I wanted to. Alongside another volunteer, I even spoke in front of 80 people in Cambridge, including the Mayor! It was terrifying, and I nearly had a panic attack, but I did it. The teenager who wouldn't talk to anyone two years before was now speaking in front of 80 people!

"Now, in early 2014, my life is better than ever. My chances of finding work are higher than I could have imagined; I have more friends, more good memories, and a more positive outlook. I have the drive to do what I want. I'm doing things because I enjoy doing them, instead of telling myself it's not worth it. I've lost weight, and have begun looking after my appearance: I wake up most days now feeling awesome. On the downside, my depression is a type that I'll never fully recover from, but Chris helped save my life, alongside those of others.

"But, more than that, she and the K9 Project gave me the drive to actually use my life. I now love my life, my friends, and my dog."

24♥

FAMILIES AND DOGS

One of our programmes – the Families and Dogs Project – is generously funded by the Kennel Club Charitable Trust – and we work with vulnerable or disadvantaged families to help them do the best for their dogs, and for themselves. When I was a social worker all those years ago, I often used to feel sorry for those dogs who, despite being much loved, allegedly, were often neglected, misunderstood, and sometimes abused. My role then was obviously to try and prevent this happening to the children in the home, and there was neither the time nor the remit to address any dog issues, unless it was one of safety for the children. Once K9 got going, I thought I would investigate whether any money was available for work such as this, and the Kennel Club – very innovatively – agreed to fund us for around four years' work.

Generally, we visited people at home, often referred by our local dog warden, RSPCA inspector, Housing Association or social worker. People also referred themselves once we'd gained some credibility and a good reputation. We offered basic advice, along the 'how to live well with your dog' philosophy, ran free, community-based training, and provided general support regarding health, training, and behaviour issues. Anything complex we referred to specialists.

The community-based dog training was always fun, as you never knew who would turn up, with what dogs, and what their level of knowledge might be. Having at least one cigarette break in the middle of the class was mandatory, and participants would sometimes turn up with tins of beer, which took some sensitive handling, as these were precisely the type of people that we wanted to reach. Not all dog trainers were interested in supporting our project, but we had a couple who stuck with it, and we ended up with a great regular group of attendees and dogs who all made terrific progress over the summer.

As the project has progressed, over the years we've noticed a difference in the type of referrals we get. Increasingly, we have been involved in re-homing dogs for a variety of reasons, and, in the current financial climate, all of the rescues are pretty full. We have worked with a number of rescues to re-home dogs, and, from what I can make out, all have had a good outcome. Occasionally, we have also found more appropriate dogs for people, and in this I have been surprised by the attitude and approach of some of the people involved in rescue work.

One example of this involved a referral by our local housing association. A lovely lady called Erica, in her 60s, with some mental health difficulties, as well as arthritis (she walked with a stick), had recently lost her parents, and her Tibetan Terrier. Erica lived in a very small bungalow with a small back garden, and had no family support. When her terrier died, and she was seriously lonely, a well-meaning friend helped her buy a working stock Border Collie pup, Jinny, whom Erica loved dearly. But (and you probably know what the 'but' is) Jinny – by now 18 months old – was making life difficult for Erica. With no training and only limited exercise, the dog was bored rigid, peeing everywhere, digging up the garden, pulling on the lead (Erica had been pulled over on more than one occasion), and was deeply frustrated.

My initial task was to try and help Erica manage Jinny, which I did, but achieved very little change. The dog was well looked after, and well fed, but other behavioural needs were not being met, which meant she was only going to become more difficult to manage. I broached the idea of re-homing Jinny with Erica, and maybe getting her a more suitable breed of dog that she could manage. Totally resistant at first, then reluctant, Erica began to consider this idea, realising that she was not meeting Jinny's needs, and maybe someone else could. To me, this demonstrated Erica's courage and honesty, as she deeply loved Jinny. But she could not provide the right home for her, and, as her health deteriorated, the situation would only worsen.

When I approached various rescues with this suggestion,

the response I got surprised me. Because Erica was asking for Jinny to be re-homed in a more suitable environment, with a more appropriate lifestyle, most of them would not entertain the idea of her giving a home to another, more suitable, animal desperately in need of one. Despite our official back-up and clear explanation of Erica's current situation, we received a flat 'no.' Erica often rang me in tears because someone at a centre had made her feel like she was the worst person in the world. Erica made a mistake in getting Jinny, and she was trying to do right by her, and to receive hurtful and disrespectful comments was not helpful, and certainly not necessary.

There was no way Erica was going to let Jinny go into kennels, so we approached the Blue Cross Charity, which provided us with Direct From Home Support. A member of staff visited Erica and Jinny, took videos, completed an assessment, and found Jinny an amazing home within a week.

Erica could not let Jinny go without another dog on the horizon, because she was terrified of the overpowering loneliness she knew she would feel. I was despairing of finding her a suitable dog when I remembered a local vet who had lots of dogs that no one else wanted (my mum had re-homed one from him a few years back). I rang him and explained the situation: did he have a suitable dog? As luck would have it, he did: Buster, a little Shitzu with a skin condition and allergies, who was a little older, liked short walks, and needed a lap and some TLC. Round he came, Erica fell in love, and Jinny went to her new home. Buster stayed, the vet provided all the medication and special skin washes he needed free of charge, and rang Erica often to make sure all was well. Buster remains with Erica, is spoilt rotten and very happy. Erica's happy, Jinny's happy, Jinny's new owner is happy, Housing Association is happy, the vet has space for another dog, and I no longer have a job to do there. What's not to like?

We have rescued many other dogs through the Families and Dogs Project, helped with the rehoming process, and assisted owners with advice and guidance to get the best relationship with their existing dogs.

Harvey, whose face appears at chapter start, broke all our hearts when we learned he had leukaemia. The temporary home we had arranged really could not care for him through this illness, and his original family did not want him back, even for a short while. Through the magic of Facebook friends, I found a lovely young vet who kindly took him in for his last few months, spoiling him thoroughly with holidays, love and care, and doggy friends. When the time came, Harvey died in her arms, after eating a usually-forbidden bar of chocolate, surrounded by love.

Duke 2 was the reason for another phone call from our local RSPCA inspector, who had been to see an elderly dog in poor condition. The inspector really felt that the dog deserved a chance, but did not think that RSPCA kennels were right for him. I visited the same day to find a poorly, thin, Whippety-cross breed dog, diarrhoea oozing from his back end, who was obviously petrified of the people in the home. His home was a bare kennel in a small, square, concrete yard. There was little water and his food was mouldy. I took him for a walk, and, pleased to be out, he brightened considerably. I felt dreadful having to leave him there, and asked around at my dog training class that evening for any ideas on how to help. Greyhound Gap rescue was mentioned, and, as soon as I sent it photos, it leapt into action.

Kev drove Duke 2 to meet a lady in Newark, who then drove him on to Stafford. He had complicated medical problems, a tumour on his rear end, an airgun pellet lodged in his skull, and other complications. He was elderly, grumpy, and a real handful. Greyhound Gap boss Lisa took him into her home, and she and her partner provided him with all the love he needed, whilst Greyhound Gap provided all the medical care. The pellet was left as it was felt it was too dangerous to remove it, but the tumour was successfully excised.

Lisa fell in love with Duke 2, and he remained with her for the eighteen months of his life, along with about seven other dogs, surrounded by love and with the best possible care. Lisa said she never would have thought that such a little dog could 'complete her life' in quite the way he did.

You can see how fabulous he looked when restored to full health in the picture gallery.

2♥5
For Duke

Today, I met you for the first time, and then I helped to end your life. It was not really something I had expected to have to do when I started the Families and Dogs Project.

Your human seemed to love you, but he hadn't made anything right for you. I was told he had no money and no way to get you to the vet. There was a van in the drive but it seemed that the person who owned it was 'not around.' There was no one else to help you. Your human owed money to the vet, and the dog warden had phoned me to ask for help. I didn't feel I could refuse.

I really wasn't sure what I would find when I first stepped into your house. There were lots of other dogs, and none of them looked great. You were lying in a corner, so very, very still upon your chair. I sensed you were close to death. I did not believe you could have become so thin in just three days, as it was claimed. I reached out to touch you and you flinched. It seemed as though you couldn't move, but I felt the pain coming from you in waves; it looked like everything hurt. I could see and feel your bones. I realised you were already travelling along your final road, but did not know how much more you would suffer if I left you there to end your journey.

So I agreed I would take you to the vet and help to end your life, and pay the bill, of course. I insisted your human came. He carried you to my car. When the vet gave you the injection he held you and said he loved you. You screamed, and I felt that my heart would break. I had to leave the room with tears falling. I felt so sad but also so full of white hot anger, and I didn't want you to sense that as you left this world.

We decide that your human can do some voluntary work instead of paying the bill. I have some ridiculously naive thought that I can maybe educate him, so that he will never again do this to a dog. I don't think he really has any idea what he has done. I don't think he feels any shame.

Duke, I so wish you had been mine, or that I had known you as a young, playful, happy boy, full of joy and life. I am so sorry you ended up with humans who were too ignorant to give you what you had a right to expect. I am sorry it was so hard for you to go at the end. Duke, I hope your pain has ended now. I do not think I will ever forget you. RIP Duke.

Postscript

Duke, I wanted to tell you that your human moved away, or so I was told, so he never did pay the bill or learn anything from us. However, the rest of the family let me help with cheap flea spray and wormers for the other dogs, and I and others worked to try and help all the dogs there.

One day I visited and there were lots of puppies. One was dying, and a man told me he had already put another 'in the bin.' There was neither food or water in the room. The mum dog – you remember Mackie – well, she was very weak and hungry. I told the man he needed to sign over the mum and the pups, and threatened him with animal cruelty laws. I was very cross. He let me take Mackie and her nine pups. We went to a vet for the night – the same one you went to – and they looked after her and managed to save all of the pups, apart from one. The nurse stayed up all night trying to save her but it was not to be. Then they all went to a rescue kennels where they were looked after really well, and got all they needed to grow round and furry and cuddly. They all found their forever homes where they would be loved, and that is due to you, Duke. Because of you I was able to help save them.

I don't think those humans are allowed to have dogs anymore, thankfully.

Duke, I've never forgotten you. Your life had an impact that you could never have known.

I just wanted you to know that.

26

LAST CHANCE SALOON

One area of work we have always grabbed as and when it came along is working with Youth Justice. As an ex-probation officer and Youth Justice Manager of many years' experience, I was really keen to see how the bond with dogs could help young people in difficulty open up, make changes, and maybe learn to trust.

For our Prevention Team (generally the younger age group of ten to 15), we create and use numerous programmes focused on having fun and thinking about our own behaviours. We provide individual programmes for local young people demonstrating anti-social behaviour to help them explore ways to change their thinking and conduct. Visits to a secure unit in a neighbouring county also occur on a regular basis, and, here, we do group workshops that focus on communication skills, stereotyping, fairness and justice, and discrimination.

We also work with young people who are either on the verge of being sent to prison, or have already been there, and are on an Intensive Supervision and Surveillance Programme. Their offences would have been either very serious, or very, very frequent – or both – for them to have got to this point. These young people can be challenging: they're often feeling aggrieved that they have to undertake 25 hours of positive activities. When, often, they are accustomed to staying in bed, and behaving exactly as they want (usually late nights and partying), to have to follow such a tight schedule is most irritating! They are usually in bed when we arrive, and do their best to make our sessions as short as possible. Sometimes, the dogs have a positive effect and they make an effort. The dogs can bring out the nurturing side in the toughest of lads, and provide a focus for discussion about so many issues that affect them. Walking outside with the dogs, and being in different places, not sitting eye-to-eye with someone, is also helpful.

The metaphor of thrown-away dogs and the grace of receiving a second chance is especially meaningful and relevant here. The image of a young man, sentenced for (accidentally) killing his stepdad, cuddling Billy with such a soft and warm look on his face will always stay with me.

My husband, Kevin, often gets to work with these older lads, because their lives are usually filled with women. Most live with their mum, and have female social workers, youth justice workers, teachers, and care workers. The lack of any appropriate male role model in their lives is very evident, and they appreciate contact with Kevin and his 'non-social work-like' approach. Kevin takes over to tell the story of Dom, who neither of us will ever forget.

DOM

"Dom was referred to us as part of the Intensive Supervision Programme. He had committed a violent attack on another man whilst under the influence of ketamine and alcohol. His background was one of deprivation and abuse, in a household where drug and alcohol misuse were the norm. Dom didn't really have anything much going for him. He had received minimal education, was very overweight, had no concept of personal hygiene, and lived in squalid surroundings.

"Initial impressions were not good. The first thing I noticed about the house in which he lived was the broken window near the front door. Broken glass littered the path. (Seven months later, when I finished working with Dom, the window still hadn't been repaired, nor the glass swept up, either.) Inside the house, the first thing I saw was dried blood all over the floor and up the kitchen cupboard doors. Dom had had a fight the previous night with his drug-user step-father.

"Dom emerged from his 'bed' on the living room sofa, closely followed by a grunting, wiggling, wheezing bundle of muscle called Hector, the Staffy. He was Dom's only real companion, and it quickly became apparent that Hector could be a strong ally in the coming weeks.

"I had decided to use Taz to work with Dom. For many young lads, the prospect of working with a powerful animal such as Taz is a big attraction. The intimidating appearance, the feeling of security, the challenge to anybody who dares mess with them is an instant hit – but, of course, being K9, we flip all of that straight on its head. I explained to Dom that Taz had 'issues' with other dogs, especially big ones, and that it was our responsibility to ensure that he and everyone in the vicinity would be safe. If we messed up, it would be Taz who paid the ultimate price. I explained that Taz's 'aggression' was actually an overly acute sense of self-preservation, brought on by having been attacked (probably many times) as a pup.

"Over the coming weeks we walked Taz in various locations. Our approach was, every time we saw a dog, we put Taz into a sit and gave him a treat. Taz's reaction to the sight of other dogs began to change, so that instead of preparing to fight, he would put himself into a sit and wait expectantly for his reward. All the time this was happening I would talk to Dom about how circumstances and surroundings affect our behaviour, and how we always have a choice. The trick is to make the right one.

"We talked at length about the situations surrounding Dom's offences, and how a combination of drugs/alcohol, company and environment was at the root of his problems. Remove one or more of the elements and the chances of encountering a problem are greatly reduced. Dom never really wanted to hurt anyone, but it had become expected of him by his peer group. He began to realise that he was a source of 'entertainment' for people who called themselves friends, and that he needed to change.

"Our work with Taz helped him to think ahead, and plan strategies to avoid potentially difficult situations. He learned how to be assertive without becoming aggressive, and he understood that, if he didn't change, there was a strong possibility he would get himself locked up, and then nobody would look after Hector. The little dog depended on Dom for his life.

"After a few sessions with Taz we began to work with Hector – the stereotypical Staffy who receives such bad press. Hector had a reputation for being aggressive toward other dogs, and Dom told me that he had taken Hector on when he discovered that the police were looking for him. The boy and the dog were so similar – I knew right away that I could make this work.

"Firstly, I impressed on Dom the importance of giving Hector the exercise he required: pent-up energy was the last thing the situation needed. The added bonus here was that

Dom also needed exercise, but the main point was that Hector should be walked after dark, when the chance of any problem encounters was considerably reduced. Dom didn't know it, but this applied equally to him. Walking Hector was a far safer option than Dom's usual 'after-dark' activities. The techniques we had worked on with Taz were also crucial. Think ahead – look ahead – have an exit strategy if things go pear-shaped. Reward good behaviour. Stay calm in stressful situations. Expect other people to do daft things, and be prepared at all times. These strategies were designed to put Dom in control; give him responsibility, and, ultimately, show him that he could effect change and do something worthwhile. As a result, Dom became a much better dog owner, and a more responsible and thoughtful person to boot. I can't think of any other way that Dom could have made these changes. He loved his dog – neither lecturing nor punishment could come anywhere near the power of that.

"I'll always remember Dom with great fondness. On our first session I told him how much I hated being asked 'what time does this finish?' within the first few seconds of a session. Dom never failed to ask me that question, with a great big smile on his face, every time I took him out. We would walk Taz, and talk a lot about making good choices, invariably ending up at the cafe, where Dom would squirt mountains of ketchup into his egg-and-bacon butty, and then proceed to spread it all over himself, his clothes, the table, and his surroundings. He always seemed a little surprised by this, and would ask "I've made a bit of a mess of this, haven't I?"

"As my work with Dom was drawing to a close, his Mum was diagnosed with cancer. She, despite her difficulties, was the only hope for a successful outcome for Dom, as nobody else gave a damn about him. I told him how important it was that he should shape up and support her, and he promised he would do his best. He promised to take care of Hector, too.

"Months later we heard the sad news that Hector had been run over and killed in the road outside of Dom's house. We don't know the circumstances, or even if Dom was at home at the time.

"Is it possible that somebody's life can be derailed when it was never on-track in the first place? Whenever I think that my life is difficult, I remember Dom."

27♥
THE F-WORD

 I have been known to use the f-word; mostly in connection with the *other* f-word. Funding. As a (fortunately) successful self-employed person, whereby my phone rings and someone asks 'Can you do (............) for us? How much will you charge? When can you start?' the idea of asking others for money is alien to me. But the experiences I have had when doing so have taught me many lessons, and caused me much frustration (another f-word) ...

In the current climate of catastrophic cuts to public spending, it is doubly difficult to apply for funding, when all around people are losing their jobs, and those who need help with their lives are struggling. There are regular accounts in the media of financial irregularities within large charities, which, ironically, seem to invest in the very sectors that cause some of the problems the charities purport to help with.

Our fund raising experience has been with small, local grant-makers, plus an occasional involvement with the Big Lottery, and, in the main, has been neither inspiring or fruitful.

Examples are –
- Attending a big central government launch for substantial new Youth Justice Funds, only to discover that the funding (millions of pounds) was awarded to just six of the 70 or so organisations that had applied! For K9 I was paying to attend out of my own pocket anyhow

- On a local level, working in partnership with a charity which retained £2000 of a grant for an evaluation event that never happened

- Wasting precious time attending meetings for projects where money was already available, but no plan or strategy in place to use it for those who needed it

- Funding withheld because lack of necessary management at the funding agency (which resulted in a three-month loss of earnings)

- Hearing that a new project of ours was not eligible for funding because it was not new, as we had previously worked with young people and dogs (although not in that setting or target group)

In addition, there's the current nauseating obsession with 'celebrity.' In order to be a successful charitable organisation it seems it's necessary for there to be a link with someone famous, or to become famous yourself! Courting publicity, turning myself into a 'star?' No, not me.

And the list goes on. Each time I hear of corruption, misappropriation; a CEO of a charitable organisation spending more on decorating his or her office than most of us earn in a year, another little piece of my heart dies.

My inspiration comes from the people and dogs we do this for and with, and, having read some of their stories, I believe you will understand why that is. Some of the best projects we run do not receive *any* funding: on the plus side this means that we do not have to go with someone else's target. However, this doesn't get the bills paid, either.

But we survive.

28♥

ALL WHO WANDER ARE NOT LOST

 I have a greetings card with this saying. It's in a lovely grey colour scale, and shows a Border Collie, tail waving in the air, purposefully trotting along a winding path that meanders through fields. I bought the card to pass on to someone who might benefit from it, but have not found anyone, yet, who needs it more than I do. It resonates with me, and supports my approach to the business aspect of the K9 Project.

Our limited company is the vehicle for getting things done, which I like to do. I keep paperwork to the legally-required minimum, and I don't believe in, or need, fancy titles. A healthy disrespect for bureaucracy, red tape, and systems is at the core of how I operate in business. I do not enjoy meetings for their own sake, and dislike it when it takes forever to make decisions. I am not good at detail, sometimes lose receipts, get in a muddle with paperwork, and drive my accountant (and myself) crazy.

Despite repeated, and no doubt sound, advice I have never written a business plan. I have begun one countless times; explored various templates, done lots of analysis, but still never actually *written* one. It seems alien to have to check out the competition (no-one else does what we do, anyway) and answer questions such as 'How much do you want to earn?' 'Where do you see your business in five years' time?' 'What do you want your turnover to be?' 'How are you going to address the issues of competition in a changing marketplace?' Yawn. Sigh.

I understand on an intellectual level that this would probably have made me a better businesswoman, and maybe the K9 Project would be more solvent, with a clear forward direction and a distinct business plan for the future. But, having completed such a plan, I probably would never have read it again, and I don't know that anyone else who would have wanted to, either, as I have no intention of borrowing money for the project. Should a rich philanthropist appear about to give us funding with which to continue our awesome work, and wants a business plan, I will write one.

Up to this point, progress has occurred naturally. We keep overheads to a minimum – we don't even have an office – and, while it would be terrific to have somewhere to show off some of our great photos and the awards we have won, this seems an entirely unnecessary expense when I have a phone and a computer at home! A flexible approach has given us an opportunity for variety, and our low overheads are probably one of the main reasons we continue to survive. By responding to requests from people we never thought we would work with, and listening to those who have attended our programmes and learnt from them, we have devised far more different approaches and programmes than I would ever have thought possible: maybe not right for everyone, and not right for some funders.

But.

Even though I may appear to be wandering, I am not at all lost.

29

LITTLE THINGS

Life is made up of many small things which have great significance for me. But it sometimes seems that the trend is for everything to get bigger, better, greater.

Let's take celebrity charity challenges, for instance, which may seem a bit random, but bear with me ... I truly admire the folks who take part in things like this: their bravery, drive, and sheer determination matched only by the intricacy and difficulty of their respective challenges. Each year, the challenges are bigger and more complex: running a marathon is no longer enough, let's make it ten marathons, or maybe a marathon every day for a year? How about we combine swimming, cycling, running, and anything else we can get in there to make the challenges so life-changing that participants are never the same again? And while we're about it, we can watch people make themselves ill, and not allow them the option to quit.

These challenges, on the surface, are very admirable and inspiring, and raise shedloads of money, which is definitely good. But just how outrageous might they become as newer, bigger, and scarier challenges are sought? And a little voice inside wants to ask whether these big – HUGE – tests of ourselves somehow devalue the small things that ordinary folk do each day for others. I was recently told by a professional fund raiser that big charities no longer want to receive £5 from little old ladies, or £1 pocket money from kids because they cost too much to process. The organisations need to attract big buck donations to cover the big bucks they spend fund raising.

It's a bit like dogs, really. We like big stories about them. 'I would be dead if it wasn't for my dog.' 'My dog saved me.' 9/11 Twin Tower search dogs. Seriously and terribly deformed dogs who don't know that they are but just get on with living life to the full: inspirational dogs who provide motivation and a role model for people about how to live their lives. Burned dogs, mistreated dogs, tortured dogs who travel hundreds of miles to find their way home to their owners.

We love it. Heck, I love it! It makes great reading, and is very inspirational. Triumph over adversity. Ability over disability. Transformational stuff indeed.

But real life is often something much smaller, quieter; humble. For most of us, it's the everyday interactions that can make a big difference. Someone letting you out of a side street, or holding open a door. Smiling at a stranger and getting the same back. Offering another your space in a queue, or seat on the bus. A five-minute conversation; a brief connection. A simple moment of genuine, unplanned human kindness.

The majority of us, and the dogs with whom we share our lives, have not experienced and lived through major trauma, horror, or life-shattering experiences. We all have our ups and downs, ebbs and flows, of course, but not all of us need dogs to save us, rescue us, or transform us into a wholly new being. Dogs can give us a reason to get up, go out, meet other people, and brighten our lives without necessarily transforming them. Dogs give us a reason to exercise, to get out in the fresh air, and reconnect to nature and the outdoors. Yes, dogs do all of these things, but mostly what I love about dogs is their dogginess, and I miss this in a very tangible way when away from them. I admire their resilience. I love their ability to live in the moment, and to focus on the task in hand.

I love what they teach me, when I am open to learning. Often this will be in quiet, peaceful moments: small things; gifts given by dogs simply being dogs; true to themselves. It may be Cassie purposefully trotting up a garden path to say hello. Taz allowing a young, nervous child to stroke him, instilling a sense of pride. Billy dancing on his back legs with someone who hasn't uttered a word in five years of going to the day centre, but who giggles delightedly at his antics. Cassie cuddling up to Christian, and taking away his loneliness for just a while. Ruby gazing into Jamie's eyes. Buck pulling someone along on a skateboard, and making them laugh. Izzy pressing herself as tight as she can against a wheelchair-user's footrest to make it easier for him to stroke her.

Mostly small things, but sometimes with extraordinary results.

30♥
LETTING GO OR GIVING UP?

I find it difficult to define where the line is between letting go and giving up. I was raised to believe that if you wanted something enough, and were prepared to work hard to make it happen, there was a good chance that it would. I believed that giving up equalled failure. I've always had a tendency to push my way through problems, with a stubborn 'I can fix this' attitude. Within my professional and personal lives, I've always been prepared to fight for what I believe to be right and just. Sitting back and doing nothing is never an option.

But as I've grown older, enthusiasm for the fight is occasionally missing: is this due to fatigue, or maybe some late-in-life wisdom? If it's fatigue, that's probably not surprising – but not a good reason to quit. But, if it's wisdom ... well, then I suppose I need to take notice. Listening to the author Eckhart Tolle describe the cycles of life, I ponder how to follow his advice, accept the downward cycles, and just let go, especially those things that might not be serving us well. Hanging on to things and feelings that should no longer have a place in our life means there is no room for new things to enter. Letting go of the old to let in the new is part of a natural rhythm, reflected in the seasons; the essence of life's ebb and flow.

I am working on letting go of so many things that it can be pretty hard keeping track of them all. My own, once-youthful energy is definitely less than it was, and I see elderly parents lose each other, their friends and family, and their ability to function independently. A previous career, gone. Friends moving away, or even dying. And we have six dogs who we will lose at some stage, too. Nothing stays the same forever, and new things (some of it good, hopefully) come knocking on the door, asking to be let in. The passion to make a difference is still there, burning as strongly as ever, although maybe in a quieter, more realistic way, now.

I continue to learn from the dogs, and I try to live in the moment, as they do. Being present, here, in each moment, without judging, or worrying about what comes next. Acceptance. Letting go *without* giving up. Dogs are such wonderful teachers in this respect.

So, where does all this leave the K9 Project, which has, at times almost taken over my life, shattered my potentially naïve illusions about people and the system, and challenged my belief in justice and fairness. I have felt sad, angry, frustrated, despondent, exhausted, and betrayed, but I have also felt inspired, energised, encouraged, moved, amazed, and very, very humble. I have experienced some highly emotional moments, made true friends, learned new things, laughed often, and, always, always, felt deeply grateful for the many people who have allowed me to share their lives. I am constantly motivated and inspired by the challenges some of us face on a daily basis; carrying on in spite of everything.

K9 continues, even though we may not know exactly how it will develop. We have lots of new ideas, fund raising schemes, plans for new projects, and improvements for existing ones. Funding bids are in, promotional material is out, new training programmes are on the horizon. The future is bright. But if the world is not as ready for us as I would like it to be, or doesn't fully appreciate the amazing work of our dogs, or the great potential for change that this project offers, then, so be it. I can live with that, too.

There are always the dogs: leading the way, loving and lovable, quirky and funny, enthusiastic and spontaneous. Sitting by my side when I am sad, waiting for me to come home, working with a child with autism, running across the fields, having a 'chasing rabbits' dream, sitting patiently, their head on a lap offering silent comort, warmly and enthusiastically greeting everyone, saying hello to strangers, and letting you look deep into their soulful eyes.

My dogs are my constant inspiration and motivation, reminding me how to give feedback without judgement; how to love without strings.

Just dogs.
Not heroes.
Or are they ...?

♥ 31
EPILOGUE

 Since completing this book many changes have occurred, and continue to happen. The K9 Project has won several awards. One for our work with young people, a regional award for work with unemployed adults, and Phil – who you read about earlier – won a regional award for his journey as an adult learner: a tremendous achievement. And Cassie, our beautiful go-to-girl, won an Heroic Hound Award at Superdogs Live. We met Ben Fogle at the London Pet Show, where he presented us with our award in front of a huge crowd. Cassie received a three-day holiday in a dog-friendly log cabin, and me, Izzy and Ruby got to go with her! She took the day in her stride, and had made new young friends before we even got into the building!

We were granted funding to run our Take the Lead programme in two different locations, which was a fabulous opportunity! Our learning disability partnership is paying for our K9 Café venue hire, which is enormously helpful. We have new projects with young people who live in hostels, and are estranged from family and family pets, and are working for a whole year with a special needs school for students aged 16-plus with a wide range of disabilities. Our learning continues to grow and develop. We held our first ever conference – The K9 Potential: dogs leading the way to emotional wellbeing – which was a huge success, with positive feedback.

And our friend, Ryan Seville, from Neighbourly.com, spent a day with us to shoot an awesome video, so far viewed by over 1400 people on YouTube. We hope that this might help secure a different type of funding. Please check out our website to see it.

There are so many other stories I could have told you if there had been room in the book. There's Kieran, for example, who is working so hard to achieve his dream of being a dog trainer. And Nigel, who came to spend a day with shelter dogs, and felt like he had 'found himself' again; so much so that an interview he attended the next day secured him a job, after two years of being unemployed. And Shona, who battled so many severe and substantial difficulties, back with us working as a volunteer.

Each week I am lucky enough to meet new people as their particular journey unfolds. Thomas, the young lad with autism, had an impromptu dog phobia session, and his family has been able to get a dog of their own, who has trained as a PAT (Pets as Therapy) dog.

In connection with this book, I have contacted some of those who used our programmes, and have learned how their lives have panned out subsequently. I'm pleased to say it is mostly good news!

Phil – sadly – lost Patsy, who had already lived far longer than was expected, and now shares his life with Sparky, a very different but equally beautiful canine companion.

We lost Ruby. She went from an apparently fit and lively dog to slipping away in just 24 hours. But that was Ruby – our 120% dog. Cassie was diagnosed with cancer, which, typical of a dog who always thought she was a person, was more a human-type cancer than a canine one. A combination of expert veterinary care, natural and homeopathic treatments, and much internet research, our gorgeous go-to-girl did wonderfully well for a year, remaining her usual, resilient, bouncy self: an inspiration and welcome companion and friend to many. After 15 months, the cancer changed, and Cassie did her last session with a school group of seven- and eight-year-olds: we lost her three weeks later. The loss of Cassie is unfathomable and life-changing, and we are grateful that her legacy lives on in the very many lives she touched. She lives forever in our hearts.

Sometimes, when I walk into a school, or the K9 Café, I can feel that she still walks beside me, trotting along, tail waving, ever-purposeful, ever-generous, full of life and love.

Thanks so much for reading my book. I never realised how vulnerable writing a book can make you, especially when writing about something so close to your heart. I now have a far greater appreciation of the effort involved.

If you have any questions about the work that we do; want to help; get involved in some way do, please, contact me via our website (www.thek9project.co.uk).

Chris Kent, K9 Project

INDEX